THE PALMETTO BOOK

UNIVERSITY PRESS OF FLORIDA

Florida A&M University, Tallahassee
Florida Atlantic University, Boca Raton
Florida Gulf Coast University, Ft. Myers
Florida International University, Miami
Florida State University, Tallahassee
New College of Florida, Sarasota
University of Central Florida, Orlando
University of Florida, Gainesville
University of North Florida, Jacksonville
University of South Florida, Tampa
University of West Florida, Pensacola

THE

Palmetto

BOOK

Histories and Mysteries
of the Cabbage Palm

————•————

JONO MILLER

University Press of Florida
Gainesville · Tallahassee · Tampa · Boca Raton
Pensacola · Orlando · Miami · Jacksonville · Ft. Myers · Sarasota

26 25 24 23 22 21   6 5 4 3 2 1

ISBN 978-0-8130-6680-6
Library of Congress Control Number: 2020938191

The University Press of Florida is the scholarly publishing agency for the State University System
of Florida, comprising Florida A&M University, Florida Atlantic University, Florida Gulf Coast
University, Florida International University, Florida State University, New College of Florida,
University of Central Florida, University of Florida, University of North Florida, University of
South Florida, and University of West Florida.

University Press of Florida
2046 NE Waldo Road
Suite 2100
Gainesville, FL 32609
http://upress.ufl.edu

CONTENTS

Introduction

This book is a collection of stories about the cabbage palm, *Sabal palmetto*. My purported goal is to get beyond the inaccuracies and knowledge gaps to a more nuanced understanding about this particular palm. But the real goal, the goal you will never know about if you don't read book introductions, is to have you look at every plant, not just the cabbage palm, and set aside the few things you think you know about it, to wonder about its past, its relationships, and its stories. John Muir observed: "When we try to pick out anything by itself, we find it hitched to everything else in the Universe."[1] That's all I did to write this book. Wonder about one plant. I picked one thing, and indeed it has led to nearly everything, and the universe.

Trees, our most persistent and conspicuous plants, are likely to be noticed and known. How well known depends on who is paying attention. Each tree species has a story. In North America the rudiments of many tree stories are known, usually because of either their dramatic success or failure—some way they affect humans. We know about the sugar maple primarily because we pour its thickened blood on pancakes. We may not know the difference between redwoods and sequoias, but we know they are big and tall and people used to drive a car through one and, hopefully guiltily, make backyard decks from the other. And we have a special fondness for fruit trees—we were taught early on about Johnny Appleseed and not to cut down cherry trees (and if you do, don't lie about it).

As with an obscure relative we've never met, we frequently know just one lingering, hallmark story about a tree: "Natives would pound the bark and put it in streams to stun fish." "The wood is so heavy it sinks." "It was deliberately introduced and then escaped cultivation and now is a big problem."

And if you go on a nature walk with a naturalist, you will likely be told one or two or three of these factoids about those plants the naturalist recognizes. Too many nature walks are an outdoor reception line, an introductory handshake followed by some pleasantries. Speed-dating the environment. As a result, most of what we know about a given tree species is typically a bouquet of aphorisms that reduce a complex plant to a few memorable talking points. There's always a backstory.

Take the story of Palm City, for instance. On January 9 1878, a brig en route from Trinidad to Spain ran aground off the east coast of Florida. It wasn't exactly a shipwreck, more of a stranding, but even if it had broken up, the outcome might have been the same because the *Providencia* was carrying an unsinkable cargo—a shipment of twenty thousand coconuts. Two local pioneers salvaged the fruits and sold them for two and a half cents apiece. Some were eaten, but many were planted, and before long the area was known as Palm City. When citizens established a post office in 1887, they learned the Palm City name was already in use elsewhere, so they went with Palm Beach,[2] a name associated with unbridled affluence, subtropical luxuriance, and, recently, a presidential retreat.

The day before the *Providencia* ran aground, Florida was flush with a dozen native palm species in nearly every conceivable locale. Majestic royal palms grew among towering cypress and smaller swamp trees laden with ferns, orchids, and bromeliads. At the other extreme, growing in the desert-like blinding-white sugar sand of the prehistoric dunes of Florida's Central Ridge, was the scrub palmetto, *Sabal etonia.* Dwarf or blue-stem palmetto, *Sabal minor,* was found in moist to wet hammocks. The uncommon and intimidating needle palm was found from central Florida into Georgia and the Carolinas. In southern Florida there were silver palms, a Miami palmetto, paurotis palms, buccaneer palms, and two thatch palms, and the rest of the peninsula may have seemed covered with either saw palmettos (*Serenoa repens*)[3] or cabbage palms, which according to the United States Department of Agriculture may be found "as a common component of such diverse communities as freshwater cypress swamps, relic inland dune ridges, and rockland pine forests, . . . at edges of marshes, . . . on dry sites such as xeric hammocks . . . in hydric hammocks . . . also a component of both temperate and subtropical hardwoods." Oh, and also "river

The stranding of the *Providencia* in 1878 helped cement Florida's association with non-native coconuts. By Eldred Clark Johnson. Courtesy of the State Archives of Florida.

hammocks and cabbage palm hammocks or palm savannahs" as well as "on severe sites such as dunes, salt flats, barrier islands, cactus thickets, and wet prairies."[4] In other words, one should not be surprised to find the cabbage palm growing just about anywhere in peninsular Florida (although I have yet to see one living in a salt flat).

Yet despite all these different native palms, the exotic coconut was already Florida's archetypal palm, the default standard-bearer of the palm family and implicit rival of the cabbage palm. That's because a decade before the *Providencia* ran aground, the Florida Legislature directed that a new state seal include a "cocoa tree," and the early renderings of the seal feature a coconut palm (and some mountains).

A century and a half later, the cabbage palm (South Carolina's palmetto) still struggles for recognition and acceptance in Florida. So, while revered in South Carolina, among trees in Florida, the most taken for granted may well be our state tree, the slightly silly-named cabbage palm, or *Sabal palmetto*. Almost a weed in some places, they seem to sprout up everywhere and are routinely overpruned, yanked out, and generally disrespected.

Even if you suspect you have never seen what Floridians call a cabbage palm and South Carolinians call a palmetto, you have. If you have seen

A detail from Florida's second state seal, circa 1903, features the legislatively mandated "cocoa tree." Courtesy of the State Archives of Florida.

either the South Carolina or Florida flags, you've seen one. If you've received a South Carolina quarter as change, you've handled one. If you've watched napalm being dropped on Forrest Gump, you've seen them. They are strategically planted along interstates at state borders to announce to southbound vacationers arriving via I-75 and I-95 that they are entering a realm where bridges no longer freeze before road surfaces. You've seen this tree.

If nothing else, it is a plant of paradoxes and contradictions. *Sabal palmetto* is the state tree of roughly 8 percent of all Americans, despite the fact that many people don't think it is a tree. South Carolinians know their state tree is a palmetto, while Floridians are quite sure palmettos are low-growing recumbent plants with saw teeth. Despite being the same *Sabal palmetto* species, the state tree of Florida is known as the cabbage palm.

If it were not a native tree, it would be deemed an invasive exotic, aggravating landscapers in urban areas as well as nature preserve managers and grove owners. The trunks are notorious for dulling chainsaws and axes, yet dead young trees quickly decompose to a state resembling shredded wheat. It can survive having all its roots cut off, but not the harvesting of a single bud. It is almost never propagated in plant nurseries, yet somehow it is one of the most popular and reliable landscape trees sold in both Florida and South Carolina. Cabbage palms are imbued with some remarkable powers: they are the tree least likely to suffer in a hurricane, yet are frequently the most severely pruned in advance of a storm. They are one of the last trees to die due to inundation, yet grow well in xeric coastal dunes of pure sand. They are simultaneously our most flammable and most fire-resistant native tree. If they even are trees. They are assumed to be short-lived but can live for centuries (two, anyway).

And they behave contrary to most trees in many ways: having greater girth when young than when old and likewise producing more shade when 15 feet tall than at 60. Oddly, each leaf (frond) is attached to the plant in two places and typically refuses to fall once it dies. These palms are far easier to transplant when old than young. As a result, developers now rely on them to add a feeling of substance to developments where the concrete is not fully cured, and, perhaps counterintuitively, one of their main virtues as landscape plants is that they grow slowly, lending themselves to relatively stable architectural statements. These palms are frequently referred to as grasses, but they are not grasses. The trunks are not usually thought of as wood, yet the oldest occupied dwelling in Volusia County, Florida, is a two-story home comprised primarily of *Sabal palmetto* logs. It is the only North American tree that is eaten with some regularity, and Florida crackers claim it as a hallmark of their vittles, while it could just as easily be served at Mar-a-Lago as an essential component of "millionaire's salad."

Sabal palmetto makes appearances in myriad realms—it's a food, it's a valued landscape plant, it's in danger, it's taking over, it's a natural history puzzle, it plays a role in the lives of many animals and other plants, it's a craft and building material, it appears regularly in popular culture, and it's a piece of U.S. history. In fact, without cabbage palms, the United States might not exist because it can be argued that South Carolina's palmettos played a pivotal role in our war for independence. They play key roles in a puzzlingly wide diversity of ecosystems from everglades tree islands to riverine hammocks—helping to support everything from black bears to ferns, fungus, and figs. Cabbage palms inspired our nation's first environmentalist and play a conspicuous, if nonspeaking, role in what some say is the best Elvis movie ever. They may hold insights to longevity—featuring individual cells that may live for centuries. And when all human life is extinguished, or even when our planet is swallowed by the sun, alien life forms may find themselves staring at an image of a pair of cabbage palms.

Upon learning of my book topic, some people ask me why anyone would want to read about cabbage palms, a polite and deferential way of asking why anyone would want to waste time writing about such obvious plants. One might just as well explore the natural history of phone poles. Although I have been subconsciously nursing some hunches about why for about three decades, it took a while before I could answer with any certainty.

Simply put, Florida would not be Florida without cabbage palms. Nor, I believe, would South Carolina (Georgia might squeak by). They are as central to both states' past and present as oaks and pines. Yet for being such an evocative and locally common plant, relatively little is known about them. They are commonly characterized as being fast-growing, but also appear to be one of our slowest-growing native trees. They seem ubiquitous in most of peninsular Florida, but may have naturally occurred in only about two-thirds of the counties of the state. They appear in two dramatically different forms, one smooth-trunked (like a broom handle, according to John Muir), the other sheathed in a lattice of persistent leaf bases, creating a basket-weave effect. Every Florida naturalist has been forced to explain the existence of these two forms to neophytes who assume they must be different species, yet there is no adopted theory that accounts for the difference. No one seems to know how long they live, or why their natural range seemed so limited when they now are growing in Arkansas—and Seattle.

John Muir, the inventor, explorer, and proto-environmentalist, was also a botanist. During his thousand-mile walk to the Gulf of Mexico after the Civil War, the deeply religious Muir was smitten by cabbage palms, a tree that he said taught him more than any priest.[5] He encountered thousands of plants, wrote glowingly about some, mentioned others, was silent about most, and, while in Florida, commented, "It was the army of cat briers that I most dreaded."[6]

We, like Muir, bestow glowing affection on some plants, ignore most, and lavish enmity (and herbicides) on those that vex us. It can be argued that every plant gets sorted into one of three categories: the useful plants, the useless plants, and the problematic plants. Cabbage palms can be placed in all three groups.

This book is a plant biography of sorts, not an ecological monograph but a collection of essays exploring the personality and contributions of one common tree—various aspects of what is known (and unknown) about it and how it interacts both with humans and the rest of the environment. This book may end up shelved with the "commodity histories" (McPhee's *Oranges,* Kurlansky's *Cod,* etc.), but it deals evenhandedly with cabbage palms as noncommodities—what wildlife they support, how long they live, etc. It could be considered a natural history, but it is equal parts cultural history. There's some personal narrative, but it is not a memoir, and there are facts, but it is not academic. It is a peregrinating ramble about a plant you didn't necessarily think you wanted to know more about, but do. It's a demonstration of how the commonplace and ordinary can be seen, upon examination, to be extraordinary. While you are welcome to sample chapters randomly, they are grouped into three topical sections—botany, ecology, and culture—although the boundaries are not rigid. Welcome to the complicated world of the cabbage palm (or palmetto).

Part I

1

---•◦•---

Quest for the Ages

How Long Can These Things Live?

On the first page of his 1942 account of Florida folklife, *Palmetto Country,* Stetson Kennedy wrote:

> An old-timer once described it in these glowing terms: "I tell you, there's no tree like the cabbage palm. It never dies of old age, and you can't see the end of it lessen you cut it down. The sun can't wither it, fire can't burn it, and moss can't cling to it. Have you seen one bend before the wind, lay all its fans out straight, and just give so's the wind can't find nothin to take hold of?"[1]

While all of that old-timer's assertions are worth running down and fact-checking, perhaps the most tantalizing is the notion that the cabbage palm never dies of old age. That begs the question, how long can they live? That seems like a relatively straightforward question, but I've been searching for an answer for a decade, and while the answer has clear implications for how we value and manage these trees, it remains elusive.

Florida's cabbage palm and South Carolina's palmetto, *Sabal palmetto,* is one of an old group of trees—the Florida Museum of Natural History has a fossil *Sabal* palm leaf from the Florida panhandle that is somewhere between 16 and 18 million years old.[2] And fossils of *Sabal magothiensis* from the Late Cretaceous period are more than 80 million years old.[3]

As a genus, then, they are survivors. And, intriguingly, some of the individual cells that compose the cabbage palm also appear to be survivors—apparently living as long as the tree itself.[4] But the longevity of cells, or the

Fossil *Sabal* palm, *Sabalites apalachicolensis,* Middle
Miocene in age, found at Alum Bluff, Apalachicola River,
Liberty County, Florida. Photo by Terry A. Lott. Florida
Museum of Natural History.

perseverance of a genus, tells us little about the lifespan of an individual
plant. With many species of temperate trees, dendrochronologists can decode seasonal variations in wood production to calculate age. Trees add
softer wood faster in the spring or beginning of the growing season and
produce denser wood in the summer or later in the growing season. The
result is annual rings that convert cross-sections of trees into continuous

data recorders that have unlocked timelines regarding droughts, fires, and even hurricanes.

More ambiguous seasons in the subtropics and tropics produce less clear growth rings, but palms present special problems. Most trees we deal with, the oaks and maples, are dicots and grow both upward and outward each year. The new wood is produced by the cambium layer, a relatively thin layer between the bark and the interior wood of the tree.

But palms represent a completely different approach to being a tree. They are monocots, and monocot species such as the cabbage palm do not add girth or annual rings as they age. The absence of annual rings thwarts the possibility of determining age simply by cutting the trees down and counting and cross-dating. Some researchers, desirous of determining the age of cabbage palms, took to counting leaf scars and then making some assumptions about how many leaves are produced each year. Aside from the challenges of counting scars on the eroding trunks, this approach is predicated on a fairly constant rate of annual growth, which cannot be assumed.

In 1970, student researcher Kyle Brown planted some cabbage palm seeds he had collected in North Carolina. While his home garden conditions in north Florida are unlikely to replicate what may have happened in the wild, Dr. Brown reports that the time for the palm to emerge and form a trunk, the establishment period, took about twenty years. Thirty-nine years later, the tree was 39 feet tall, conveniently yielding an average rate of growth of 1 foot per year. Yet the vast majority of this growth took place in the second half of the time.[5]

Scott Zona and Katherine Maidman, working at Fairchild Tropical Botanic Garden near Miami, set out to determine not how long various species of palms live, but how fast they grow. They found a twenty-fold difference in growth rates between the slowest and fastest species. They had only one cabbage palm for which they knew the complete age. It had been a seed forty-two years earlier. We might expect from Dr. Brown's experience that the palm would be roughly 42 feet high. Actually the palm measured 263 centimeters, or 8.6 feet tall. Since they measured to where the lowest live leaf attaches to the trunk, additional height, possibly as much as 6 or 8 feet, should probably be added to approximate the total height. Still, that's about four and a half inches per year, although it's probable that, like the Kyle Brown tree, the majority of the tree's height developed in recent years.[6]

The Fairchild Gardens tree was planted about fifty years ago and, curiously, may be the oldest cabbage palm for which we know with certainty its complete age. Aside from the Kyle Brown and Fairchild trees, there are few other cabbage palms of a known age, because so few people plant them from seed and then keep track of the seedlings as they grow.

By the time someone notices a young cabbage palm in his or her lot, it may be one or more decades old. The explanation lies in the curious manner of cabbage palm growth, which involves a lengthy establishment phase. Aside from being a hidden, underground process, the establishment phase can last years, and, consequently, many observers fail to account for this stage of growth. As a result, researchers resort to a variety of techniques to estimate how long it takes to grow from seed to a tree with an aboveground trunk. The results are startling. In one study area the fastest cabbage palm developed an aboveground trunk in about fourteen years, but it was calculated that only half the seedlings would have a trunk after fifty-nine years.[7]

So whereas a fifty-nine-year-old live oak is likely to be a big tree, in the Waccasassa Bay area of Florida, a fifty-nine-year-old cabbage palm is, on average, just getting to the point where it has an aboveground trunk. Researchers concede that the establishment phase may be significantly faster when the tree is grown in more favorable conditions. One report states that ideal nursery conditions can produce a cabbage palm with an aboveground stem in only seven years.[8] But the fact remains that cabbage palms in the wild spend far longer in a vulnerable juvenile phase than most would imagine.

Once the cabbage palm emerges from the ground with a true trunk, it enters a phase of relatively rapid growth. Abetted by an extensive root system, the palm can shoot upward, creating the impression that its average rate of growth is swift—no doubt contributing to the popular perception that they are both fast-growing and quite likely short-lived.

In 1860, the year John D. Rockefeller entered the oil business, Elizabeth Cady Stanton was campaigning for women's suffrage, and American housewives were advised to cook tomatoes for at least three hours before they could be safely eaten.[9] On January 23 of that year, a government land office surveyor was working west of Lake Okeechobee near Fisheating Creek, and he needed to mark a section corner at the northwest corner of Section 30, Township 40 South, Range 32 East. In addition to placing a monument at corners, surveyors were instructed to find "witness trees" and measure the distance and bearing from these trees to the corner. If the main corner monument was lost or removed, the location could be reconstructed by measuring back from the witness trees. Searching for a witness tree, the surveyor found a cabbage palm and marked it. Surveyors subsequently found that same witness tree at least twice more, once in 1938 and again during a resurvey related to the Fisheating Creek sovereign lands controversy in 1999, when the palm had two smaller, presumably offspring, trees associated with it. Surveyor George "Chappy" Young was so impressed with the persistent palm that he wrote an article about it and kept track of the tree. When the palm died of unknown causes, possibly lightning, it was at least 139 years old, and probably significantly older. Touched by the loss, Chappy wrote a subsequent article for a surveyor's journal entitled "Death of a South Florida Surveying Icon (this is not what you think)."[10]

Were Chappy Young's "Sable [sic] Queen" growing along Waccasassa Bay, we might justifiably estimate an age somewhere around two hundred years, by counting the establishment phase necessary for the palm to achieve an aboveground trunk. A two-hundred-year-old cabbage palm would be stereotype shattering. Many people have trouble believing an organism with the diameter of a light pole could live for a single century, much less two. Many sources simply duck the lifespan question. The Florida Native Plant Society simply offers "long-lived perennial."[11] The Smithsonian Marine Station at Fort Pierce features a cabbage palm profile webpage that includes a section with the helpful, if redundant, heading "Age—Size—Lifespan." Beneath the heading are descriptions of stem diameter, leaves, fruits, and roots. There is no mention of either age or lifespan.[12]

On a rainy Saturday in December 2009 my wife, Julie, and I left Sarasota and headed east past Myakka River State Park in search of a witness tree. About halfway to Arcadia we turned south onto an unpaved ranch road and made our way to Buster Longino's elevated home tucked in an oak palm hammock at the end of Tampensi Trail, a woods road named for a native orchid.[13] We swapped stories with Buster's wife, Jane, while tea brewed and light rain tapered off. Buster produced a slim folder with facsimiles of an early surveyor's notes related to the Longino Ranch. In January 1843, the same year that Charles Dickens's *A Christmas Carol* was published,[14] and before Florida became a state, Henry Washington was surveying township lines in that region. Working south along a line that

The taller palm in this image was first marked as a survey witness tree in 1860 and was subsequently relocated in 1938 and 1999. Photo by Chappy Young.

would become the eastern boundary of Manatee County, Washington found himself in the midst of a lightning-ridden, fire-prone landscape, which burned every few years and was dominated by wet and dry prairie. As he entered section 24 he encountered a sawgrass pond and worked his way south through sawgrass, savanna, cabbage palms, and live oak. He set a half-mile post, and since no flammable wooden post could be expected to persist for many years, he chose two witness trees in order to improve the chances of another surveyor finding the same half-mile point in the future. The only significant nearby trees happened to be cabbage palms tall enough to receive a distinguishing scar. Since cabbage palms survive fire, they would have to do. One stood to the northeast about 6 feet away and was 14 inches in diameter; the other measured 2 inches smaller, about 25 feet to the southwest.[15]

We shoved off in Buster's Honda Pilot and passed the old Ranch House, where a cluster of housebound hunters waited for the skies to clear. Buster pulled over to chat, and we asked the assembled group how long they thought cabbage palms might live. "Not long around here," one wag offered, an oblique reference to the ranch's intermittent enterprise that allows some cabbage palms to be dug for the landscape trade. In a more serious vein, someone speculated twenty years. Implicit in his estimate of twenty years was a mind-set that these are not substantial plants, likely not trees at all, but some large vegetable, a woody stalk of celery perhaps, that shoots up, declines, and decays in a generation. I wondered how his estimation of these trees might change if he thought they might be older than the massive live oaks we stood beneath.

Back in the Pilot we worked our way up the section line that forms the northern boundary of the ranch and headed east to the eastern township line, passing through two gates and dodging several cows and their three-week-old calves.

After months of drought the recent rain had altered both the landscape and Buster's planned route. We passed a hammock and turned to hug one edge as we worked our way south, skirting a now-drained wetland. Anticipating deep holes ahead, Buster left the woods road and found a spot to cross a drainage ditch using a combination of Buster's dead reckoning and memory. We parked and walked down the township line as Buster checked each fence post looking for some distinctive flagging he had placed some time ago to remind himself of the half-mile post when he personally chained his east boundary and discovered the bearing tree.

Spotting a substantial post with faded ribbon, Buster turned and plunged into the adjacent palm and oak hammock, ominously commenting on how long it had taken him to find it the last time he located it. Within seconds he was pulling aside some low fronds from a small palmetto to reveal a clean trunk with three distinctive double hatchet marks. Julie measured the circumference of the tree, which worked out to a diameter of slightly less than 12.5 inches. The dense hammock made it hard to measure the height, but we finally determined it to be slightly less than 39 feet to the lowest living leaf. Could this possibly be a tree that grew only about 35 feet or less in 166 years? That works out to roughly 2.5 inches a year. If so, it begs the question of how old a 90-foot-tall cabbage palm might be. More than 250 years?

Having determined the height and diameter, we measured back to the marked fence post, and my heart sank. It was more than 70 feet away, while the original survey notes indicated the farther of the two witness trees was less than 25 feet from the half-mile post.

My mind raced through possibilities. Were we in the wrong section? Might there be another section with a marked palm at a more plausible distance? Desperate, I volunteered to walk a mile south to another half-section point. While I walked south, Julie and Buster took the vehicle around and waited at the south end of a large wetland straddling the township line. Meeting them, and still unsuccessful, I slid into the back seat, sat back, and puzzled.

Buster turned the vehicle and stopped. He peered out the driver's window at a 12-foot-high pine. "I wonder what killed that pine?" he mused. The pine had invaded the wetland as the result of drought or drainage, and I assumed it had simply been killed by the return of water. But two smaller live pines appeared behind it, deeper into the wetland. I looked more closely, saw that the lower bark was charred, and suggested fire. In addition to having a forestry degree from the University of Florida, Buster has been dealing with fire and pines professionally longer than I have been alive. He drove off slowly, contemplating the demise of a single pine.

Buster's father and grandfather started the ranch in 1934, primarily as a turpentine and timber operation. Buster continued the timber operation and gradually added other income streams: cattle, citrus, hunting leases, palm removal leases, and the latest—gopher tortoise mitigation. Three hundred acres of the ranch are now receiving tortoises relocated from development sites such as the new solar photovoltaic array in De Soto County. As a result of disparate strands of the ranch that require attention, any change that could impact trees, cows, grapefruit, quail, deer, turkey, cabbage palms, or gopher tortoises is a matter of both personal and professional interest to Buster.

As we worked our way back to Buster's, he picked a different route and stopped to comment on a dead cabbage palm. He had noticed it previously and was planning to have someone come take a sample to make sure it wasn't diseased. It's humbling and reassuring to travel with a man who knows more about the soils, water movement, plants, and animals on 8,000 acres than most suburbanites know about their lots.

Once back at his place I checked the surveyor's field notes and reluctantly concluded that the witness tree we located and measured was unlikely to have been marked by Henry Washington, since it was way too far from the half-mile post in the fence line.

Driving back to Sarasota, Julie mentioned that Buster had told her of a second abandoned line of fence posts that reflected a historic calculation of the township line, one that had been subsequently corrected and moved eastward. Hope seeped back in.

Had we been seduced by the newer fence line and Buster's flagged post, taking it for Henry Washington's when it might be an entirely new line? Now my mind raced back to our entry into the hammock. Buster had plowed in, briefly commenting on another palm wrapped with orange flagging and sporting a bright surveyor's tag nailed into the trunk. We had more or less ignored it in our quest for the 1843 tree. Was it a newer witness tree, or might it mark a spot about 25 feet from the tree we had sought?

Once home I shot an email to Barry Wharton, an expert in reconstructing historic habitat patterns in the Florida landscape. His unusual career could be called either environmental historian or historical ecologist. Barry uses multiple tools to deduce a landscape's backstory, and early surveyor's notes are one of the most important. Barry replied as I expected he might. He pointed out that the three chopped scars near the half-mile mark did not comply with the "General Instructions" given to Government Land Office surveyors.[16] Barry recommended I contact "an old land surveyor."[17]

Some Floridians are sensitive about being deemed old, but I sensed Chappy Young accepted his role as a respected and venerable surveyor interested in both the history of surveying in Florida and Florida itself. Chappy's return email confirmed what I was learning the hard way: "There is a lot more to proving out a witness tree than meets the eye." He went on to take away hope, noting: "All of the marks I have seen on cabbage trees that were proven to be original, were worn down to just a shallow dish-shaped surface and not recognizable as axe marks at all." But then he blew on the embers by suggesting, "However, the more protected they are from wind and weather, the better the marks will remain.[18]

Here is what is known about the Longino Section 24 cabbage palm: It is in the vicinity of a half-mile marker set by Henry Washington in 1843. The tree has three sets of distinct horizontal axe marks, and they have been there at least since the mid- to late 1950s, when Buster located them. The marks are widely spaced and do not represent a half-hearted attempt to chop the tree down. The tree is not visible from the fence-line road and is in a fairly dense hammock. If it could be found living, would Buster Longino's Section 24 bearing tree be the oldest-known living cabbage palm? Possibly, but more work is needed.

Are there other candidates? Yes, Barry Wharton suggested he knew of other cabbage palm witness trees to check out,[19] but I never ran those down. Instead I started focusing up the east coast of Florida on Fort George Island, where a tantalizing double row of cabbage palms lines Palmetto Avenue leading to the Kingsley Plantation.

In order to embrace those cabbage palms as the oldest known, we have to accept the account of Henry Nehrling, one of many botanical explorers drawn to Florida. One-quarter of his 1933 book *The Plant World in Florida* is consumed with descriptions of palms, and he devotes a portion of his description of cabbage palms to Palmetto Avenue on Fort George Island.

According to Nehrling, in 1813 an overseer named Mr. Houston directed Mr. Kingsley's slaves to plant a double row of cabbage palms for one-third of a mile along the road approaching the plantation. Then Kingsley returned and "discharged" Houston, since Kingsley thought the palm planting was "useless work."[20] Cabbage palms are notoriously hard to transplant when young, by which people mean small. It is hard to find anyone who recommends moving them with less than 8 to 10 feet of trunk, a condition that can take decades to develop. I believe the palms (evenly spaced 20 feet apart) along Palmetto Avenue are the originals and therefore actually stretch back into the early eighteenth century. To walk along this avenue, now surrounded by dense hammock, and imagine the effort involved for slaves to plant hundreds of trees by hand is humbling. And contemplating the fact that the remaining live palms may be the very trees planted by slaves can make one feel quite nearly reverential.

Several problems vex the Nehrling account, which involved his retelling of a story over a century old. The first is that the 1813 date does not lend itself to corroboration. While there is general agreement that slaves planted the trees, there is controversy over when. Mr. Houston was probably John Houston Mcintosh, who owned Fort George Island before Zephaniah Kingsley. According to a Florida Archives blog posting, Mcintosh left Fort George Island in 1813, and Kingsley didn't secure a mortgage until 1817, suggesting there may have been some ambiguous period during which the two disagreed about property management.[21] Mary Kingsley Sammis, a daughter of Kingsley who was born in 1811, remembered helping her mother plant palms,[22] but since the length of the palm planting is now two-thirds of a mile, she could have been working on an extension of the original planting. So while the original plantings could have been as late as

The hundreds of cabbage palms planted by slaves along Palmetto Avenue on Fort George Island are almost certainly two centuries old. Note: The engraver made the fronds pinnate. Engraving from *A Winter at Fort George, Florida,* Fort George Island Company (Boston: Frank Wood, ca. 1888), 8. Courtesy of the State Archives of Florida.

the 1830s, it is safe to conclude that some of the remaining palms are more than two hundred years old, when one accounts for the years needed to grow a cabbage palm to a size that will withstand transplanting.

The other, and more conceptually problematic, but less likely issue is the suggestion I heard at the National Park site that the current trees are replacements, a belief that is probably the result of unknown observer's stereotyping the species with a presumed short lifespan, but one that is difficult to conclusively refute. Exactly how does one prove substitute trees were not planted? In order to be alive today and conform to the modest lifespan that supposedly would have doomed the originals, the replacements would have to have been planted in the twentieth century. Field measurements taken of the existing trees suggest there were originally at least 296 palms lining the avenue, and my wife and I counted 125 that were still alive in 2009 and more that were still standing but dead. While it seems logical that the magnitude of effort needed to manually dig and replant three hundred palms would have been noticed and noted, the absence of such reports does not definitively rule out substitution. The ten dozen cabbage palms still lining Palmetto Avenue are probably the oldest continuously documentable cabbage palms known, but at this time they lack the conclusive proof of continuity that may be afforded by trees marked by Florida's early surveyors.

About two hundred miles up the coast from Fort George Island lies what is now known as Sullivan's Island. In June 1776 summer had descended on what was then called O'Sullivan's Island, a South Carolina barrier island that for nearly seventy years had been known primarily as a quarantine station for arriving slaves.[23] Now its role was changing, and the oppressive heat, mosquitoes, and palpable presence of British warships made it an unfortunate yet crucial time to be engaged in heavy labor. Patriots had been struggling for three months to construct a modest fort in a position that commanded the Charleston Bar, an area of shoals that limited access to Charleston Harbor. The main British fleet had marshaled outside the harbor at the beginning of the month, and it was clear the British were reconnoitering a strategy to get past the slowly materializing fort and take Charleston.[24]

Unfortunately, sandy O'Sullivan's Island had no rock, and the only trees of any measure were contorted live oak, dense wax myrtle shrubs, and scrawny palmetto trees, many no bigger around than a pie plate.[25] The oak was extremely tough and had already been recognized as possessing significant military importance in ship construction. On Ossabaw Island, 140 miles down the coast, trunks and massive branches of live oak were being cut to form the now-arcane futtocks, hanging knees, hawespieces, breasthooks, stanchions, knight heads, and bow timbers necessary to build wooden warships.[26]

Yet O'Sullivan's Island live oak was not well suited to quickly constructing fortifications. Roughly twice as dense as cedar, the tough trunks were very hard to cut, heavy to move, and the branches seldom straight. The landscape, desperation, and necessity dictated that any fort on the island was to be constructed of the only likely materials in abundance—beach sand and palm logs. The palms were a fairly constant diameter, typically straight, and in good supply. They stacked easily. The Patriot Council of Safety described specifications for the palmetto logs. They were to be "not less than ten inches diameter in the middle, one-third to be eighteen feet long, the other two-thirds twenty feet long."[27]

Some of the logs came from O'Sullivan's Island itself, and some were brought "from other sea islands and the mainland."[28] As the English attack became more imminent, one easily imagines an abandoned front of palm-felling where at some point during the third week of June workmen metaphorically dropped their axes to man the fort in preparation for the impending onslaught.

For backwoods Carolinian Patriots new to the American coastal South, these unlikely columnar palm trees must have seemed novel indeed. They were dealing with tree trunks that get thinner as they age, with leaves that are attached to the tree in two places instead of one, and a tree that appears in two different forms—one encumbered by a lattice of remnant leaf bases called bootjacks, and another so smooth that naturalist John Muir would later compare them to broom handles.[29]

On the morning of June 28, 1776, Thomas Jefferson was preparing to share his final draft of the Declaration of Independence in Philadelphia as four British warships arrived about 400 yards offshore O'Sullivan's Island with the intention of subduing the partially constructed fort so they could enter into Charleston Harbor unimpeded. As the British reported after the battle: "The Piemeto [sic] Tree, of a springy substance is used in framing the Parapet and the interstices filled with sand. We have found by experience that this construction will resist the heaviest fire."[30] The fort held, and the British ships suffered extensive damage and withdrew.

The following day, the British set one of their grounded ships afire and abandoned it. When it exploded, Colonel Moultrie reported seeing "a grand pillar of smoke, which soon expanded itself at the top, and to appearance, formed the figure of a palmetto tree."[31] What the twentieth and twenty-first century know as a mushroom cloud was a palmetto tree cloud to eighteenth-century South Carolinians.

Five days later fifty-six men in Philadelphia signed Jefferson's draft. Perhaps if Francis Scott Key had been at Fort Moultrie at the start of the Revolutionary War instead of at Fort McHenry during the War of 1812, we'd be opening baseball games with someone singing about palm logs and sand and a plume of smoke that looked like a palmetto instead of "the ramp parts we washed" and the "donzerly light."[32] The British failure to subdue the Patriot fortification meant Charleston was spared a British takeover until March 1780.[33]

The original Fort Sullivan (renamed Fort Moultrie) was subsequently lost to a storm and replaced by a series of more ambitious fortifications using more substantial materials while the fortuitous palm species that secured the early victory found its way onto the South Carolina state flag during the Civil War when the secessionist state needed an evocative symbol of desperate, yet successful, military defiance.

Perhaps somewhere today on Sullivan's Island, behind an aging cottage a few blocks from the beach, there might be a lone cabbage palm, with a shallow dish-shaped surface, the eroded trace of an abandoned axe bite taken at the dawn of our nation.

Or perhaps a road crew clearing a fallen palm on Palmetto Avenue on Fort George Island will take a break and, absentmindedly kicking a decomposing palm root mass, reveal a long-lost, tarnished coin from the early 1800s. A technological breakthrough in aging wood samples might obviate the need to know the date a palm was planted in order to determine age. Or perhaps curious and dedicated surveyors and environmental historians will track down another survey witness tree that pushes the oldest documentable cabbage palm back past the two-century mark.

I'm confident the cabbage palm entering its third century is out there, resolutely ignorant and apathetic regarding our desire to quantify and

The first Fort Moultrie was constructed of palmetto logs. George R. Hall, *Defence of Fort Moultrie, S.C.,* engraving, based on a painting by Johannes Adam Simon Oertel, 1858. The Miriam and Ira D. Wallach Division of Art, Prints and Photographs: Picture Collection, The New York Public Library. New York Public Library Digital Collections. http://digitalcollections.nypl. org/items/510d47e0-f561-a3d9-e040-e00a18064a99

document. Its eventual discovery, documentation, and notoriety will force a reevaluation of our cavalier treatment of these trees, which we have collectively come to believe are short-lived. And its discovery will reawaken our awareness of the ways in which today's landscapes link us to the past. And what if Stetson Kennedy's old-timer is right? What if patient natural historians someday confirm that cabbage palms never die of old age? Then we open ourselves to the possibility that there may be living native palms that have silently and patiently witnessed virtually all of what we invaders think of as southeastern American history.

2

—•—

Are Cabbage Palms Trees?
Wood? Grass?

Are cabbage palms trees? Most people regard answering this question as a trivial, no-brainer situation comparable to Supreme Court Justice Potter Stewart contemplating obscenity in *Jacobellis v. Ohio* ("I know it when I see it").[1] Even if people may have trouble providing a definition of a tree, they seem to intuitively know whether a cabbage palm is or is not a tree. Unfortunately, their responses tend to settle on either "obviously" or "of course not."

I once helped with a field botany course that involved teaching students how to identify plants, and the standard academic approach (instead of paging through picture books) involves making sequential choices using a plant key. If your plant in hand is found in the plant key you are consulting, then successfully making a series of choices should lead you to the scientific name for that plant. It's like playing twenty questions except instead of answering a question with a yes or no, one is forced to choose between two (sometimes three) competing descriptions. The plant key poses a choice, and you look at the plant in hand and make a decision. The questions frequently start out easy (*Plants with woody stems?* or *Plants with herbaceous stems?*) and get more detailed and arcane as one proceeds (*Leaves and stems scurfy, grayish throughout* or *Leaves and stems without scurfy indumentum?*). Such choices require either a working knowledge of botanical terms or a handy copy of something like Harrington and Durrell's *How to Identify Plants,* which, through descriptions and sketches, explains the meanings of terms such as antrorse, biturbinate, caespitose, and didymous.[2] The goal of the keying process is to arrive at a single agreed-upon scientific name consisting of a genus and a species. Palms are in the Palm plant family. The cabbage palm genus is *Sabal,* and the species is *palmetto.*

But even without obscure botanical terms, the easy questions aren't always easy calls, and to demonstrate we asked the botany students to form groups of five or six and agree on a definition for the common term "shrub." We overheard them talking about plant heights, trunk diameters, and the number of trunks. And after a half hour, none of the groups had definitions they completely agreed upon. If we had asked what a tree was, we would have had similar results. And had we asked the students if, for instance, oaks were trees, they all probably would have agreed. Personally, I favor performance-based definitions: If the plant in question can pass the following tests, I am willing to confer tree status: If you could fall out of one, and you call it a tree, and if it lives longer than you are likely to, then it is totally appropriate to call it a tree.

Then we could have taken them out in the field to encounter what is commonly known as running or runner oak, *Quercus pumila.* It usually looks like a groundcover or a dense, low-growing shrub. You won't learn much about it in Gil Nelson's *Trees of Florida* because "the running oak is actually a shrub" that "seldom exceeds one meter in height and is not considered a tree for the purpose of this book."[3] So even though all oaks are closely related, not all oaks are considered trees.

This illustrates the fact that a plant's taxonomic status—that is, what plant family it is in—cannot always predict whether or not all the members of the family will be considered trees.

It is the Linnaean system of nomenclature with its plant families based on relationships that leads to the most common cabbage palm misunderstanding: the belief that cabbage palms (and palms in general) are giant grasses, like bamboos, (which actually are grasses and can achieve towering forms). Here's an example: *Tampa Bay Times* gardening columnist Penny Carnathan argued: "Palms are monocots, in the same family as corn, grass and bamboo. Having one as our state tree bothers me in the same way a misspelled word does."[4] Of course, having a gardening columnist who doesn't know palms are not in the same family as corn, grass, and bamboo bothers me a lot more than a misspelled word does.

Confusion may be traced to the very short shrift now afforded to botany in schools, with the result that most people have a fairly shallow

Areca palms (*left*) look superficially similar to bamboo (*right*), but note that the bamboo has branch buds (circles) at the nodes, whereas the palm has no buds at the leaf scars. Illustration by the author.

understanding of how plants are related to each other. And, to be fair, some palms look superficially similar to some bamboos. *Chamaedorea* and areca palms, for instance, are commonly called "bamboo palms" because their stems are reminiscent of bamboo culms. So if you have been laboring under the impression that palms are grasses, that's perfectly understandable. Not your fault. Please support better school funding.

Grasses are a distinct family of plants (the Grass family), which is technically called the Poaceae (with about ten thousand species).[5] Palms, then, are in their own completely different plant family, the Palm family, which used to be called the Palmae, but now is called the Arecaceae (with about 2,600 species).[6] When people assert that palms are grasses, they are attempting to point out that palms and grasses are both monocots, flowering plants in the same subclass, the Monocotyledonae, as opposed to the dicots, the subclass that contains trees such as oaks. Other monocot families include bananas (Musaceae), gingers (Zingiberaceae), orchids

(Orchidaceae), onions (Amaryllidaceae), bromeliads (Bromeliaceae), and lilies (Liliaceae). No one goes about claiming onions are orchids, or pineapples are bananas, but the assertion that palms are grasses is a well-entrenched misunderstanding of plant families.

While calling palms grasses may be considered a forgivable mistake, there is a more ominous and demeaning variant that I commonly hear and one that should be challenged immediately—the derogatory assertion that palms "are *just* grasses." This is the vegetable equivalent of dismissing someone as just a housewife, or just a student. What do you mean "just a grass"? Show some respect. If you or anyone you care about has ever eaten any of the cereals (wheat, rice, corn, rye, barley, millet, and oats), then you have benefitted from "just" grasses. The seeds from these grasses "arguably provide more food energy worldwide than any other type of crop."[7] And cereal grains are only part of the grass family—don't overlook the contributions of bamboo and cane sugar. So if there is any plant family that humans would be totally lost without, it is no doubt the grasses. First runner-up actually might be the Palm family due to the importance of coconuts, dates, and (perhaps regrettably) palm oil.

So cabbage palms are not grasses, but are they trees? That depends on whom you ask because different groups of people use varying definitions.

Trees, it turns out, are not a level of Linnaean taxonomic organization. While there is a Palm family, there is no tree family or tree order. The orders and families are about relationships, not roles. But it is worth thinking about roles. Plants take varying forms that can be considered strategies for survival. Many grasses grow low and horizontally (which is why we use them for lawns). Vines rely on other structure for support as they seek sunlight. Epiphytes grow on other plants. And trees use strong, woody material to become perennial, self-supporting organisms that can tower over most of those other plant forms. These observations about plant form led Danish botanist Christen Raunkiaer to devise an alternative way of grouping plants that was based on their form (actually the location of their meristems) rather than on their relationships.[8] It is not that Linnaean taxonomy is better or worse than Raunkiaer's categorization—they are both valid and serve different purposes. But one cannot necessarily draw direct connections between the two systems of organization.

Raunkiaer's category phanerophytes includes shrubs and trees—perennial plants with buds (meristems) well aboveground. What we would call trees, Raunkiaer would call meso- or mega-phanerophytes (which may help explain why his system isn't better known). Despite the unfamiliarity of his naming system, you can see how it is helpful to be able to group plants based on what you can see of their form rather than on their relationships, which are not always obvious. The Palm family, for instance, contains several different plant growth forms: trees, shrubs, and lianas. Lianas are woody vines—rattan furniture is made from palm lianas.

Implicit in questions about form are aspects of permanence. Trees are long-lived. We can't imagine a tree that lives only for one year because the challenges of marshalling enough woodiness to have buds well aboveground would be daunting.

Field botanists can't immediately see plant relationships, so people trying to identify plants in the field invariably view plants such as cabbage palms as trees—because in order to start keying a plant out, you need to describe it, and the descriptions are almost invariably based on what is visible, even if you need a hand lens. In Long and Lakela's *A Flora of Tropical Florida,* the *Sabal* (genus) key asks users to choose between *Shrubs with stems or rootstocks subterranean,* or *Tree, rootstock perpendicular.*[9] If you don't choose the second "tree" option, you will never get to the cabbage palm.

Consequently, many of Florida's most renowned historic botanists and plant explorers have routinely considered cabbage palms to be trees. They include F. Andre Michaux,[10] John Kunkel Small,[11] William Bartram,[12] Henry Nehrling,[13] and John C. Gifford.[14] Many botanists and naturalists of the late twentieth century also refer to cabbage palms as trees. In addition to Long and Lakela, they include Moyroud,[15] Little,[16] Craighead Sr.,[17] Sibley[18] and Ward & Ing.[19] So field botanists and naturalists consider cabbage palms to be trees because they adopt an arborescent (treelike) form— they are Raunkiaer's meso-phanerophytes. Therefore, based on form, from the perspective of field botanists, cabbage palms have to be trees. And that explains why a cabbage palm appears as the cover photo of Nelson's *Trees of Florida.*[20] Case closed. Except.

The authoritative Marie Selby Botanical Gardens published a blog entry featuring cabbage palms (tree of the month) that included this introductory phrase, "Although not technically a tree,"[21] so I contacted the director of research at Selby, Dr. Bruce Holst, who explained that contrary to the approach taken by field botanists, taxonomic botanists (or "systematists")

choose to define trees and wood as coming only from dicots, so, by definition, that leaves palms out. They might just as well say: "We don't care how
woody or hard a palm trunk is, we are never going to call it wood, and,
by the way and by definition, no monocot can be considered a tree." So,
field botanists think cabbage palms are trees, whereas taxonomic botanists
don't.

Another approach to the tree question uses a functional test to define a
tree, based on how people use it. One way trees serve us is to yield wood.
Using Long and Lakela's plant key again, the path to the palms involves the
following choice: *Plant with woody stems* or *Plant with herbaceous stems?*[22]
If you don't choose the first woody option, you will never get to the palms.
So field botanists believe palms are woody, but being "woody" doesn't necessarily settle the question of whether cabbage palms yield wood, just as
being friendly doesn't necessarily make one a friend.

The Raunkiaer categories may be seen as reflecting differing plant strategies—trees are self-supporting life-forms that have strategies to grow above
other competing plants, where they can access sunlight, whereas vines, lianas, and epiphytes depend on others for support. In order to tower over
other plants, stiffness is required. And when plant stems or trunks get stiff
enough to support towering growth (requiring a saw or axe rather than
a machete to cut one down), we think of those stems or trunks as being
woody. It is certainly true that palm trees don't look like a lot of other
plants we call trees. (But have you noticed that we call them palm trees?)
Palm trees are just another solution to the question, how can this perennial plant be substantial enough to get its leaves up above other competing
plants? That calls for woodiness. Woodiness, I argue, constitutes a prereq-
uisite for any possible consideration of being wood.

Yet you can't head to a lumberyard and buy a cabbage palm two-by-four.
So foresters and taxonomic botanists will say "no"—monocots can't produce true wood. Never mind bamboo or coconut palmwood flooring—it
may be as hard as wood and used as if it were wood, but it is not wood.

However, if your rule-of-thumb definition of wood is long-lasting plant
material that can be used to build log structures, then cabbage palms meet
the criterion. Dead and felled they find service as decorative tikis. I've
found homes and other structures around Florida that use palm logs not
only as posts supporting porches but also as walls. The palm log cabins at
Myakka River State Park were built by the CCC in the 1930s and, recently

rehabbed, have held up well enough that the park can charge seventy dollars per night for one. Some committed wood turners work with cabbage palm, and John Mascoll has created beautiful cabbage palm vases. Cabbage palms have been used for corduroy roads and both bridge supports and dock pilings, where they had a reputation for not rotting.

So when I am asked whether I think cabbage palms are trees or are made of wood, I resort to the same practical performance approach I used with shrubs. If you wouldn't attempt to cut it down with a machete, and when you cut it down you end up with logs, and one can build a house, or most of a house, out of those logs, then whatever it is made of, I'm willing to call that wood.

As for you, you can call *Sabal palmetto* a cabbage palm or a palmetto and a tree or not. Whatever your choice, there's a botanist who will support you. If it pleases you to think they consist of wood, go right ahead; there's an argument to be made. Just don't call them grasses. They're not, and there's no botanist around who will come to your aid.

3

———•———

Bootjacks

Fronds, How They Work (and Fail)

I view myself as an old-school natural historian, a patient observer combining qualitative insights with prolonged rumination. After more than ten years, I imagine I've looked intently at far more cabbage palms than all but a few botanists (and nurserymen), and sometimes it takes me a while to develop an insight.

First, before any insights, some definitions and cabbage palm anatomy. Cabbage palms don't have branches or twigs. What they do have is leaves. Big leaves. There are three conspicuous parts of a cabbage palm leaf—the blade, the petiole, and the colloquial bootjack (leaf base) where the leaf attaches to the trunk or stem.

The leaf blade is the big pleated fan at the end of the leaf. The structure of the palmetto leaf blade is different than two more common palm leaf structures. Coconut and date palms sport feathery pinnate leaves, whereas the iconic *Washingtonia* palms you might associate with Palm Springs or Los Angeles have palmate leaves. Palmate leaves have leaf segments that originate from a common point, sort of like the way your fingers radiate from your palm. The pinnate (feathery) leaflets occur parallel all along either side of what is called a rachis, the continuation of the leaf stem, or petiole.

Cabbage palms and several other palms possess leaves that are similar to palmate leaves, but which are called costapalmate. Superficially they may seem palmate, but instead of emanating from what is called a rachis on pinnate palms, the leaf segments originate from a structure called a

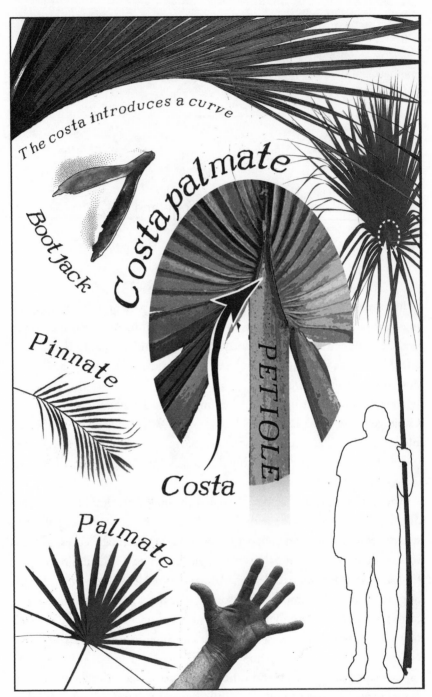

The costa introduces a curve

Costapalmate

Boot Jack

Pinnate

PETIOLE

Costa

Palmate

Cabbage palm fronds are neither palmate nor pinnate. The curving spear-shaped costa extends the area from which the leaf segments emanate. Shade-grown cabbage palms can have very long petioles. Illustration by the author.

costa, which is a tapering, curved extension of the petiole into the blade. So, instead of all leaf segments originating from a small area as they do on palmate palms, the long downward-curving costa adds dozens more segments so the blade doesn't look like a flat two-dimensional pleated fan but, rather, an arcing, three-dimensional flounce. The top surface of the costa resembles a spear point and typically features a small green excrescence known as a hastula.

The petiole, commonly called a leaf stem, is the structure that connects the leaf blade to the trunk. At the very center of the first palm heart I dissected were the tips of tiny folded leaf segments. Then the minuscule end of a costa, sporting some tiny folded leaf segments on either side, then entire leaf petioles, and finally the bifurcated petioles, which deserve explanation.

When one finds a wishbone-shaped, trunk-hugging base of a palmetto leaf on the ground, its size and dramatic "Y" shape distinguish it from all dicot leaves. These fallen leaf bases are called bootjacks (frequently shortened to "boots") because they figuratively look like old-fashioned bootjacks and they can actually function as bootjacks. One merely stands on the singular base of the "Y" with one foot and lodges the heel of the other boot in the crotch of the "Y" and pulls to remove the boot. Bootjacks undo what shoehorns do. Dry palm bootjacks can really get a campfire roaring, but aside from removing boots and some awkward craft projects, there is not much to be done with them.

Still, that "Y" configuration seems to ask its own question: why are cabbage palm leaves attached to a plant in two places?

If you have a working knowledge of how a leaf is attached to a dicot such as a maple or oak, then, unfortunately, you're starting out with a handicap when it comes to understanding how a cabbage palm leaf is related to the rest of the plant.

The answer lies in the arcane architecture of the cabbage palm. So asking other questions may be more revealing. How can a single leaf on any tree support the weight of an adult human without failing? How do these large—gigantic, really—wind-daring leaves remain on the tree in hurricanes when dicots can be stripped of their leaves? The fact is that the living palmetto leaves aren't really attached to the trunk in any common understanding of the word. We usually think of common (dicotyledonous) leaves as attaching to a twig via a leaf stem, called a petiole. Most dicot leaves are modest in size, and it is hard to think of a petiole any heftier than coat-hanger wire. Bigger leaves do have heftier petioles—magnolias, say—but most nontropical petioles don't have to do a lot of work or persist for years.

Cabbage palms face different challenges: their leaves are enormous when compared to oaks or maples, and each one is precious. For a mature dicot tree, losing thirty leaves is trivial. For a cabbage palm, that could easily make it leafless. So the palm needs an architecture that can reliably secure these giant leaves. Instead of being attached at one point to a twig (remember, palms don't have twigs), the leaves can be thought of as lopsided tubes—cylinders that the rest of the tree grows up and through. So imagine a series of nested tubes, one inside the other, like an old-fashioned car radio antenna or a modern selfie stick. Each tube has a leaf attached to it on one side. And in addition to that leaf being a tube that completely encircles the tree, it is secured in place by the older tubes outside it. The result is a leaf no human could possibly pull off of the tree. I can stand on one leaf next to the trunk, and it remains firmly attached.

As the palm grows, each leaf tube has to expand as the tubes inside increase their diameter. Most of the tube apparently deforms or stretches as the trunk expands, but the base of each leaf is torn asunder, resulting in the Y-shaped bootjack. Zona: "The petiole sheaths the stem for a short distance, and as the stem expands, the persistent petiole base splits longitudinally forming a characteristic crisscross pattern."[1]

Here's an observation: all manner of epiphytes grow quite happily in the bootjacks of older, taller cabbage palms, but one seldom sees the same diversity near eye level. Why is that? Eventually I realized that each frond goes through a predictable sequence of changing angles of attachment. All the fronds start as a central, vertical "spear leaf" that is then pushed aside by subsequent leaves. On the younger palms the fronds resolutely point upward, even the dead fronds. These lower fronds typically clasp the trunk tightly and leave little room for any natural detritus (soil in the making) to accumulate. Not only are the fronds on the older, taller palms more relaxed and open to collecting organic material in their leaf axils, but the higher fronds are closer together, more bunched up. I've noticed that when one contemplates a tall cabbage palm, the leaf scars near the ground are far more widely spaced than the leaf scars immediately under the canopy. Apparently the additional incremental growth between leaves is diminished on taller trees. This effect is frequently conspicuous in the appearance

Occasionally bootjacks appear to line up in eight rows, reflecting the fact that it takes an average of eight fronds to encircle the trunk. Illustration by the author.

of trunks that have not lost their bootjacks where the bootjacks appear packed together. One result is that, unlike younger palms with their erect fronds, the lower fronds on taller palms point downward, creating the spherical canopy associated with unpruned cabbage palms.

So the arrangement of leaves on the cabbage palm varies, which changes the appearance of the tree. What governs the arrangement of the fronds? It is obviously to the tree's advantage to deploy leaves in three-dimensional space in such a way that minimizes one leaf shading another. That involves spacing the leaves both around the trunk and vertically. Each leaf is separated from its nearest neighbor by being a certain distance higher and rotated a certain amount around from it.

How many leaves does it take to encircle a cabbage palm? Sometimes eight. And if you're observant you can find cabbage palms that take exactly eight, creating the appearance of eight vertical rows of fronds going up the trunk, each row separated from the next by 45 degrees (since 45 × 8 = 360). This arrangement of leaves is called octastichous or eight-ranked or

⅜ phyllotaxy. On these uncommon palms that sport eight vertical rows of leaves, one might assume that each subsequent leaf diverges from its immediate older predecessor leaf by 45 degrees. But that's not how it works.

Yes, theoretically each leaf could be offset 45 degrees from the one below it. That's the approach taken by *Pandanus* plants as each new leaf appears almost immediately above the previous leaf, like a living spiral staircase. The helical arrangement of the narrow, strap-shaped leaves yields the common name of *Pandanus:* screwpine. But cabbage palms don't have narrow leaves, and having one big fan leaf follow almost on top of the previous leaf, even if separated by 45 degrees, would occlude a lot of sunlight, in part because it would be only one incremental height unit above the preceding leaf.

In reality, each frond is rotated approximately 135 degrees (or three 45-degree increments) from its immediate predecessor frond, so it is nowhere near it. So, after eight exactly 135-degree passes, the subsequent ninth leaf appears directly over the starting leaf and is eight incremental leaf height units above the starting leaf.

As noted, only rarely do the fronds diverge by exactly 135 degrees and therefore appear to line up vertically.

Now, all these various angles are of no concern to the palm, which grows as it will, based on its genetic programming and the specific environmental site conditions. And, since the angles of leaf attachment change through time and space, botanists rely on features other than these angles to distinguish different palm species, and consequently leaf angles and arrangements are not subject to much meticulous scrutiny.

Except by one group of people: digital artists. Digital palm trees are needed to populate architectural renderings, video games, and CGI (computer-generated imagery) movies. These palms have to appear somewhat realistically in a three-dimensional digital world. The challenge for digital designers is to create an accurate mathematical description of a palm that can be viewed from any angle or modified to depict different height trees, etc. There are some very credible digital date palms available. There are passable coconut palms. But as of the summer of 2020, no one has produced a convincing cabbage palm. There may be a simple explanation. Date and coconut palms are found worldwide and constitute a global market. Cabbage palms are a niche market affecting just a handful of states. And why would anyone populate a fantastic video game with cabbage

Top: A cross section of the bud reveals each leaf base originates approximately 135 degrees from its predecessor. *Bottom*: Bootjacks are the palm frond leaf bases. Palms with retained leaf bases are referred to as being booted, palms lacking boots as "slick." Illustration by the author.

palms if more iconic and exotic date and coconut palms were available? So there probably hasn't been the financial incentive for teams to spend hours of fieldwork cutting up palms, making hundreds of distance and angular measurements to create mathematical equations that describe how cabbage palm leaves are deployed. And if and when some team gets around to it, they will have to create two variants of cabbage palms—one with remnant leaf bases, and one without.

John Muir commented that the trunks of cabbage palms were as "smooth as if turned on a lathe,"[2] whereas Harriet Beecher Stowe, writing in *Palmetto Leaves,* described the same species of tree as having "scaly trunks"[3] and "a regular criss-cross of basket-work."[4] How can that be? How can some cabbage palms appear downright armed and dangerous, sporting jagged spears creating a spiky lattice effect that projects several feet from the trunk, whereas others could be mistaken for an innocuous phone pole?

These two different forms (and, for the dedicated observer, a puzzling array of intermediate types) seem so different from each other that many people not only suspect that they are different species, but some people actually call the clean-trunked examples *Sabal* palms and the spiky variety, cabbage palms. But they are, nonetheless, all the same species of tree. These forms can appear growing side by side, and it is not always easy to convince casual observers that these two versions are two incarnations of the same plant.

A few things are not in dispute. Really tall cabbage palms are invariably clean-trunked ("slick"), with the exception of the recently deceased leaves up near the green canopy. And young, typically short, palms are almost always, but not quite always, booted.

The spiky latticework is the result of two fairly uncommon phenomena in the plant world: first, as noted, the pattern of leaf attachment on the trunk seems far more complicated and interesting than that found on many plants. Second, the dead leaves are persistent—they hang around on the plant for a while after they are dead. And while O. Henry's protagonist Johnsy and obsessive homeowners in *New Yorker* cartoons wait for the last leaf to fall, in general most trees deliberately and neatly shed their dead leaves in a timely manner. In fact, a number of other palms do as well.

Plants lose dead leaves in two manners. Most, but not all, dicots shed their leaves through a process controlled by the living plant. Monocots seem more inclined to retain dead leaves, a condition known as marcescence. The marcescent leaves hang around on the plant until processes not governed by the living plant remove them. These include rotting, fire, wind, animals, and human activity.

For a conventional dicotyledonous tree, such as a maple, shedding leaves is a necessity. If dead leaves simply hung on year after year, they would be weighing down the tree with useless tatters that block sunlight and consequently reduce the new leaves' potential for photosynthesis. But since dead palm leaves are always at the bottom of a healthy palm canopy, there is no photosynthetic price to pay by having a few hang about.

When the sugar maples, sumacs, birches, and hickories conspire to attract leaf-peeping tourists to New England, they do two things with their leaves: pull out useful compounds and prepare to jettison the leaf. The process of the maple leaf separating from the twig is analogous to a space shuttle leaving a space station—hatches must be sealed before separating. In order to effect a neat separation, the leaf forms a corky bulkhead called an abscission layer. The abscission layer ensures that when the leaf detaches, the tree's vital fluids don't leak out of an open wound. The cabbage palm pulls nutrients, particularly potassium, out of the dying leaf,[5] but it skips the whole abscission layer part of the process—the leaf simply dies in place.

And while I can show you multiple contrary examples, most cabbage palms less than 15 feet tall are covered with persistent leaf bases. Conversely, as noted, really tall cabbage palms have trunks free of bootjacks, except up at the base of the canopy.

The puzzle is this: if the palms are not actively controlling or mediating their leaf loss, what accounts for some trees having clean trunks while others remain covered in dead leaf bases? Most authors either avoid the subject entirely or include decidedly passive descriptions with no causal explanation. The USDA Forest Service blandly states: "On many trees the leaf bases or 'boots' remain securely attached while on others they slough off, leaving a fairly smooth trunk."[6] John Kunkel Small offered two causes: "First, *time;* being softer than the trunk they will eventually decay and fall off. Second, *fire;* successive fires will remove them much quicker than the normal disintegration of the tissues."[7]

Palm experts Tim Broschat and Monica Elliot also offered two explanations:

In sabal palms (Sabal palmetto), people often ask why some trunks are smooth and others have an attractive pattern of old leaf bases ("boots") firmly attached. There are two explanations for this. The first is that the natural retention versus shedding of old leaf bases by individual palms is probably genetically determined. Since all sabal palms are seed propagated and thus are genetically different, some will retain their old leaf bases for 50 years or longer, while others shed their old bases after 5–10 years. The second explanation is that those leaf bases that may remain on a sabal palm are manually cut off by the installer prior to transplanting into the landscape, making the trunk appear smooth.[8]

Addressing their explanations in reverse order, it is true that many transplanted palms have been dressed with chainsaws creating an appearance not unlike a hunk of gyro meat, which appears to be a cylinder approximated by repeated downward straight-sided shaving. That explanation is both pervasive and persuasive in the urban landscape, but not in the hammocks and pastures from which those palms came.

I am open to the possibility that there could easily be a genetic component that helps explain some of the marcescence. Some trees may simply have fronds made of tougher stuff due to some genetic programming. In addition, favorable growing conditions could lead to some trees having more robust, tougher, and therefore more persistent leaves. Or conversely, perhaps rapid growth that takes place after establishment is less substantial. However, I've found several examples with bootjacks at the top and bottom of the trunk with a bare midriff and the opposite: trees with clean trunks above and below with a band of marcescent leaf bases in the middle. And why would we often see remnant boots at the base of the palm that are probably decades older than the ones lost above? These examples suggest to me that there is not a single genetic predisposition that establishes a uniform leaf-losing strategy or "personality" for each tree but, rather, that localized conditions along the trunk can vary.

In any case, I don't think cabbage palms actually "shed" their marcescent leaves so much as lose them to forces they do not control. That view seems to be reflected in Scott Zona's monograph on the genus *Sabal:* "Over time, the remains of the petiole base fall or rot away, but while present they provide a foothold for epiphytes and hemiepiphytes, as well as a home for insects and other small animals."[9]

When the fronds are green and living they are resilient, but once dead they become dry and brittle. Based on my observations of palms not subjected to human pruning, the dead leaves typically fail in two ways. The most spectacular and sudden is that they are consumed by fire—high-speed oxidation. Rarely the entire leaf, from the leaf base on the tree to the end of the frond, falls away from the trunk as one unit. But more commonly the giant leaf blade and part of the petiole get caught by the wind and snap off, anywhere from a few inches to several feet from the trunk, leaving the bootjacks and protruding leaf stems on the tree and collectively creating that distinctive spiky effect. And here's a funky fact: the trunk end of the petioles of the blown-off fronds typically points in the direction from which the wind came.

It would appear that both of these leaf-failure variants are governed by physics—at some point the loads on the terminal leaf blade overwhelm the strength of the increasingly brittle petiole. This suggests that the balance between load and petiole strength or attachment strength can change through time. A single living leaf that will easily support a 180-pound human when living, eventually just snaps off. Higher loads will lead to dead leaf loss and so will weakened petioles or attachment. In its most extreme form the load can be applied by a human, but natural processes such as high winds can strip dead leaves as well. And while I have no experimental data, I'm convinced that whereas the loss of the dried leaf blades and petioles are governed by physics, the loss of the bootjack where it is attached to the trunk is mediated by biological processes that the palm does not initiate or control. And I believe it is these biological processes that result in weakened attachment to the trunk.

If you inspect cabbage palms relentlessly and study the persistent leaf bases, those bootjacks, you'll start encountering examples with holes in them. My observations of the holes have led me to conclude that these are typically made by our smaller woodpeckers, particularly the downy woodpecker, and my contention is they are not simply tentative practice holes but represent deliberate forays to extract edible invertebrates. At least six different species of woodpeckers have been found exploring cabbage palms, some on the trunk and some on the petioles.

Downy woodpeckers excavate dry bootjacks in search of insects. Photo by the author.

And if you take a knife to old remnant bootjacks that are still on the tree, you'll frequently find lacunae or galleries, voids in the leaf bases. Fresh green fronds have no such voids. So my hypothesis is that small boring (excavating, not dull) organisms, no doubt insects and possibly other phyla, excavate the interior of the leaf bases and weaken the attachment points to the trunk, setting the stage for detachment. I have been surprised by native bees that excavated homes in the petioles as well as Florida carpenter ants, locally known as bull ants,[10] making their nest in the decomposing leaf bases below the canopy. Buddy Mills, who has harvested lots of swamp cabbage, told me these bull ants are common in the decomposing dead palm fronds below the live fronds. Ant biologist Jack Longino confirmed that what Floridians call "bull ants" are a common carpenter ant species: *Camponotus floridanus.* Dr. Longino added that it "wouldn't surprise me at all if they weakened bracts and hastened their shedding from trees."[11] Our largest common woodpecker, the pileated, prefers carpenter ants.[12] A University of Florida fact sheet confirms these ants can be found "under old leaf petioles in palms" and that they "seek either existing voids in which to nest or excavate only soft materials such as rotten or pithy wood and Styrofoam."[13]

It may be possible to document the weakening of leaf attachment by insects. I contacted an entomology class at New College of Florida in the hope they could conduct some straightforward research and ultimately provide a persuasive story explaining why two specimens of the same tree, growing side by side, can look so different. A student, Iliana Moore, emerged and volunteered to look closely at 110 different cabbage palms on the campus. Her observations contributed to the suspicion that the actions of the Florida carpenter ant may be a primary factor weakening the attachment of the marcescent leaves.[14]

In addition to insects compromising the integrity of the leaf bases by removing material, it is also probable that fungi are weakening the attachment to the trunk, by decomposing the leaf bases. The more I started looking for fungi, the more I found—the lead suspect is a white polypore, possibly a *Wrightoporia*.[15] I'm developing some facility at recognizing palm trunks that might sport the white fruiting bodies. It seems most common in older leaf bases, and I suspect that rainwater flowing down the trunk (called stemflow) creates moist conditions that favor fungal growth. Iliana Moore also came to suspect fungi as organisms that weaken petiole attachment.[16]

Curiously, and perhaps counterintuitively, I have observed that this weakening and detachment frequently starts higher up on the trunk and then proceeds downward. This finding eliminates the possibility that loss of the bootjacks is based on the age of the leaf, since the remnant lower bootjacks could easily be a decade, or more, older than the higher, fugitive bootjacks. I don't have a great hypothesis about why that may be, other than to say the earlier, lower fronds may be of tougher stuff. It may be that the older leaves closer to the ground grew more slowly and substantially, and some higher "teenage-growth-spurt" leaves are simply not as robust. At any rate, I notice that some palms still carrying their dead fronds get a decidedly congested look as the angle of the upward-pointing fronds relaxes and they start losing their grip.

In April 2015, we adopted a dog, and I commenced daily walks around the block. I noticed a booted palm that seemed as though it was about to lose its boots, so I started photographing it with my cell phone, attempting to stand in the same spot each time. I assumed that in a few weeks I'd have a bad version of time-lapse photography showing the transition from booted to slick. Five years later, the palm still has most of its boots. So, contrary to my prior assumption, the loss of bootjacks can be a protracted process measured in years.

If the remnant leaf attachments are weakened by either age or insects, then all that remains is for a raccoon or a windstorm to initiate a cascade in which the trunk goes from enrobed to shorn in a relatively short period of time. That's because I suspect the weight of the loosened, failing fronds above exerts a significant gravitational imperative on the bootjacks below. And while palm petioles don't seem very dense, rain-soaked, decomposing leaf bases could amplify a cascade in which the incremental added weight of each detached frond above increases the likelihood of lower frond failure. Indeed, it is not uncommon to encounter a palm surrounded by numerous fallen boots—indicating that they all fell at about the same time, possibly in one event, rather than in some orderly, one-at-a-time chronological sequence based on their age.

Marcescent leaves are usually not considered aesthetic features, and if palm poopers had to coalesce on a single complaint about cabbage palms, it is probably those brown, droopy, persistent dead leaves.

So why are all those dead leaves hanging about? We could simply conclude "it is what it is"—just the ways things are, how God made them, or how they happened to evolve, and that they play no known role in the life of the palm. Marcescence is sometimes viewed as a primitive trait, as if marcescent plants are simply not sophisticated enough to do things right. But it is intriguing to speculate on possible advantages that might be conferred by having dead leaves figuratively and literally hanging around.

It could be that creating abscission layers is demanding, and the plant is conserving resources by not doing so. Or the dead leaves could create conditions that favor other organisms such as insects, bats, or epiphytic plants that somehow benefit the palm—some form of symbiotic relationship. But the theoretical function that intrigues me most has to do with the possibility that retaining dead, dry leaf bases might help eliminate competition from other trees.

Dicots have a living cambium layer immediately below the epidermis or bark, and that makes the tree vulnerable to heat, cold, girdling, or any stress that can penetrate the bark. As a result, many dicots have trouble surviving fires unless they possess special adaptations such as thick, insulating bark.

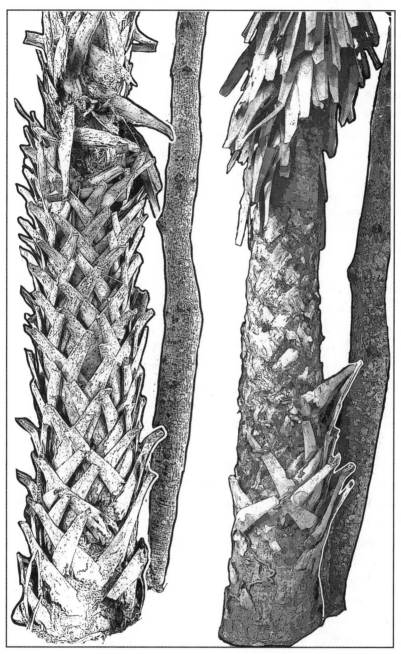

Bootjack loss occurs more slowly than many might imagine. This is the same palm captured from a similar vantage point 1,720 days apart (April 5, 2015, to December 20, 2019). Illustration by the author.

On the other hand, mature cabbage palms, monocots, can famously withstand fire. The accumulated fuel represented by the dry fronds can ignite as a huge fireball, but so long as the growing tip, the apical meristem, is protected, which it almost invariably is, the tree will survive and throw out new growth. Cabbage palms are found in a number of plant communities that are considered fire-dependent. In fact, a mature cabbage palm with a full complement of dead boots is probably simultaneously and paradoxically the most flammable and most fireproof tree in the forest. What if the persistent, marcescent flammable leaves actually conferred a competitive advantage by stockpiling fuel that enables the palm to generate intense heat levels that scorches nearby plants? I believe I've found an example of a cabbage palm wounding (but not killing) a slash pine, which has its own fire defenses.

Today all one sees is a tall slash pine a few feet from a clean-trunked cabbage palm. But the base of the pine facing the palm is wounded and charred. Slash pines are another fire-adapted species with bark that can tolerate moderate amounts of fire, so something intense seems to have happened here. The wound seems too low to be a turpentine scar, and the dark char suggests that fire at least was a contributing, and probably the causal, factor. There is no evidence of char on the palm. I try to rewind the years in my mind to a time when the palm was young and resplendent in dry, dead, flammable bootjacks. The clean palm trunk today is only a couple of feet from the pine, which means that at one time the leaf bases were very close to, if not touching, the pine. If the fire that charred the pine happened then, the burning leaf bases would have created intense heat adjacent to the pine. We'll never know what happened with these two trees. But I've seen a similar pairing with a cabbage palm and an adjacent fire-damaged live oak.

When I drive I-75 in North Port or just east of the toll booth at the west end of Alligator Alley, it's easy to spot dead, charred pines towering over rebounding, verdant palms—apparent evidence of a pine forest transitioning to a cabbage palm hammock as a result of a severe fire. Researcher David Fox: "Because Sabal palmetto populations are not devastated by wildfire, some interesting fire effects result where cabbage palms are abundant but not the dominant overstory species. This author has found many locations where a recent wildfire killed dominant pine or hardwood stands and left a nearly pure stand of under- and mid-story sized cabbage palms to thrive with no overstory competition."[17] Fox goes on to wonder "if some of the

The pine on the left was probably fire-scarred when the burning bootjacks scorched the trunk.

pure cabbage palm stands that exist today may be the result of a stand-changing wildfire that occurred a century or more previous."[18] And noted conservation biologist Reed Noss wrote, "Many observers have noted the ability of the fire-adapted cabbage palms to 'throw' flaming fronds ('fire-brands') into the oak canopy killing or pruning oaks to the palm's advantage."[19] This damage to the oak's canopy, even if the rest of the tree remains alive, no doubt increases the amount of sunlight reaching the rebounding palm canopy. And that, we may hypothesize, improves the prospects for the palm.

The retention of dead fronds and dead leaf bases (bootjacks) is a common, but typically unexplained, feature for numerous palm species and a common cabbage palm complaint, but the striking appearance and ecological function of retained bootjacks in terms of fire ecology and the species they support is so different from most trees in southeastern forests that palm bootjacks should demand more attention, appreciation (and research) than they have received so far.

4

———•———

Hunting Island

The Curious Recurved Seedling

There's no better place to understand palmetto seedling anatomy (and barrier beach behavior) than South Carolina. The best-known coastal islands are the most intensively developed ones that can be reached by car: Hilton Head, Kiawah, Sullivan's, Isle of Palms, Fripp, Edisto, and Pawley's Islands, and the long stretch anchored by Myrtle Beach. Here you will find people ever-attentive about storms and erosion and no shortage of rock revetments, golf courses, and fine dining establishments. At the other end of the spectrum are the islands primarily in their native state, and these islands' names are less well-known: Bay Point, St. Phillips, Capers (both the southern and northern), Pritchards, Hunting, Pockoy, and Bull. Some are large, some quite small. Some government-owned, some private. What they have in common, aside from being relatively untouched, are their bleached and weather-beaten trunks of pines, oaks, and palms and the fact that their forests are yielding to the Atlantic Ocean.

Travel to South Carolina's favorite state park, Hunting Island. Like many barrier islands, Hunting Island is far longer than it is wide (by about eight times), and like virtually all barrier islands, it is constantly changing.

We pulled into the entrance station with what I assumed was an unusual request: "We're trying to get to the section of the island where the palms are falling in." The attendant volunteered that could be a number of places but that we should head up north toward the lighthouse. The one-lane asphalt park road threads through a dense pine and palm forest. While there were scattered oaks, I gawked at the unlikely mix of towering pines and palms. We parked and walked east, drawn by the sound of the Atlantic and a vista of silhouettes of lower palms with an overstory of taller pines. We emerged on the Atlantic beach just south of a rock groin, which no doubt had been placed there in an effort to reduce erosion near the lighthouse.

North of the groin, erosion was rampant, but rather than seeming depressing, or foreboding, the area was attracting people: photographers, dog walkers, couples, and at least one palm researcher. Some surf-battered palms lay horizontal on the beach, resting before the next tide. From the air these look like gigantic swizzle sticks—long and straight with a bulbous end formed by the root mass. Others were still standing, waiting their turn on a half-eroded bluff. Farther up the beach there were palm logs that had washed over the beach and been deposited inland. This went on for a third of a mile.

An eroding bluff creates an opportunity to informally study normally hidden, subterranean palm anatomy, especially the strange and curious "saxophone." This unanticipated structure shatters common understanding of how tree roots look and behave. The notion that tree roots could have handedness or sides is a foreign concept. But the belowground portions of developing cabbage palms actually have two discernibly different sides. We assume that roots radiate outward from trees in all directions in appropriately dendritic patterns, and they do in mature palms. But on the eroding bank I found an exposed young palm that had fallen, entangled in the roots of another tree. As I blew away the sand, beautiful overlapping scales, surprisingly like colorful turkey feathers, clasped the bulbous bottom of what is known as the "saxophone." Exposed to sunlight, this normally hidden portion of the young tree had responded with photosynthetic zeal, belying its connection to the emergent leaves. This half of the saxophone is really a subterranean portion of the leaf-bearing trunk. Folded back against itself, in true saxophone fashion, is the portion that bears the roots.

But let's start at the beginning. In order to comprehend how different the cabbage palm strategy is from that of a conventional (dicotyledonous) oak or maple, let's review how a young oak comes into existence. When that squirrel-buried acorn germinates, it starts with a trunk like a thin wire topped with multiple leaves. That initial oak trunk, however wiry and puny, is visible from the first, and it adds incremental girth each year, like making a hand-dipped candle by dipping a wick in molten wax, with each new layer adding to the girth. As the tree grows upward (primary growth), its trunk grows outward (secondary growth). Each year it adds wood to its circumference, enabling the support and metabolic transport system of the tree to increase apace. This one-year-at-a-time approach allows the trunk to start small and grow proportionally more substantial with each passing year—sort of a pay-as-you-go approach.

Unlike the incremental, pay-as-you-go strategy of the oak, the cabbage palm has to make an immense down payment at the front end, because unlike oaks and maples, there is no secondary growth of the stem—no noticeable taper marks a cabbage palm trunk, which is why they are sometimes compared to broom handles and telephone poles—cabbage palms are more or less a constant diameter from the ground up. While there are some limited exceptions to the rule that palms do not get fatter with age, it is wondrous to contemplate how that can possibly be so. How could a tree, or any plant for that matter, emerge from the ground with a trunk as thick as it will ever be? Is it even possible that when a cabbage palm trunk emerges from the ground, it does so with the circumference of the mature plant?

A cabbage palm seed—hard, black, and about the size of a rounded-off pencil eraser—finds itself on the ground. Some unknown process, a rainstorm perhaps, results in it being covered with a little soil and, in response to some set of triggering conditions, it germinates, sending up a single innocuous green shoot, nearly indistinguishable from a blade of grass. That one-leaf strategy is in marked contrast to that of something like an oak. If you find one of these grass stage leaves and dig it up, you'll find the seed is almost 2 inches deep. How did it get there? Does the seed pull itself downward into the soil? The answer, according to the man who wrote the book on the structural biology of palms, P. B. Tomlinson, is yes; the cabbage palm, and several other species of palm, have a strategy that enables them to push downward into the earth. More on that later.

Another daunting mental challenge is to imagine how a seed smaller than a coffee bean, just below the surface, can somehow transmogrify into a base of support the diameter of a medium pizza pan that might ultimately anchor a six-story-tall organism?

The inescapable, but baffling, reality is that when the cabbage palm trunk emerges from the ground, it does so with its full lifetime diameter. Restated: the palm trunk emerges from the ground with all the girth it will ever have. (In fact, cabbage palms typically lose girth as they age because physical and biological forces remove the bootjacks and then erode the outside of the trunk.) The tiny palm seedling somehow has to accumulate a massive, adult-sized trunk with modest juvenile leaves that are just a fraction of the size of adult leaves. This process involves the seedling simultaneously growing upward and downward, forming a recurved, belowground base that involves what has been called the "saxophone."[1]

The nascent trunk actually grows down into the soil while gaining sustenance from emergent leaves that are associated with the "trunkless" phase.[2] Newer, larger leaves emerge in the center, pushing older leaves outward as the tree, its trunk still hidden, gains mass. The belowground portion is establishing a root system capable of physically and metabolically supporting an entire tree, while only a dozen leaves may be visible aboveground. This crucial phase of the cabbage palm's life history is very poorly understood, especially in light of the tree's ubiquity and commercial importance.[3]

Let's consider the oak's roots. Young oak roots are roughly analogous to their branches—they start out narrow and expand in girth with age. Branches, of course, have no difficulty displacing the atmosphere as they grow, a trivial task no more difficult than walking through air. But tree roots have to displace the lithosphere—they need to physically shove soil particles aside to make room for the expanding diameter of the roots. These ever-expanding roots explain why dicot trees can lift sidewalks, or driveways. At their origin, the subterranean portion of the trunk, the roots coalesce to form the belowground portion of the tree. As a result, we're all familiar with graphic illustrations suggesting that the root structure of a dicot tree is an approximate mirror image of the branch structure. That's an inaccurate affectation of graphic designers that might lead someone to suppose cabbage palm roots extend only a few feet from the trunk. In reality they might extend as far as 50 feet.

In size and color, cabbage palm roots remind me of those crispy chow mein noodles—I don't think I've ever seen one larger in diameter than a wooden pencil, and most are the size of an iPhone charging cable or smaller. As a consequence, all the roots emerge from the rounded belowground base of the tree, Medusa-like. So instead of an upside-down branching structure, imagine an upside-down hemisphere with hundreds of roughly equal diameter wiry roots protruding from it.

To create this underground hemisphere the young cabbage palm spends much more time growing down than up. It does unfurl some leaves above

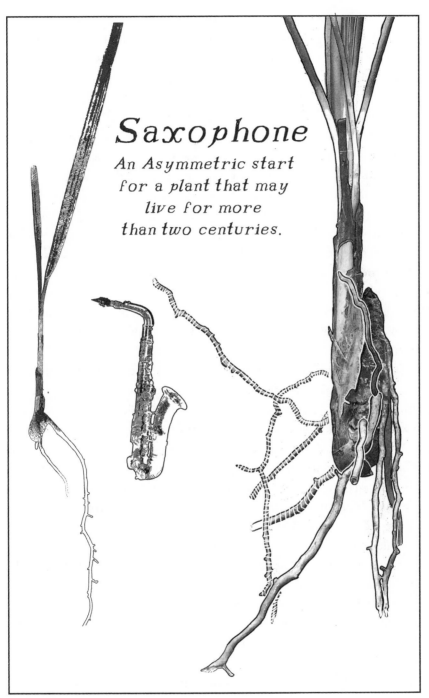

Saxophone

An Asymmetric start for a plant that may live for more than two centuries.

The asymmetry of the young underground cabbage palm has been called the "saxophone." Illustration by the author.

the soil surface, and these slowly provide the photosynthesized resources to create the bulging base of the plant pushing down into the soil. And push it does, sometimes 3 feet deep with roots extending dozens of feet outward.

The young palm has to do several things simultaneously: (1) remain anchored in the soil, (2) send new leaves upward, (3) serve as a point of origin for the roots, and (4) allow the plant's base to expand and force its way deeper in the earth. We intuitively recognize that simply applying pressure to the ground is unlikely to make it yield, hence the metaphorical futility of pounding sand. When humans want to run something long and linear (like a cable) through the ground we excavate, install, and backfill. But plants have to do their root running without shovels. Imagine your task is to bury your elbow in the soil while leaving your hand aboveground. And imagine your elbow had to keep swelling underground until it was the diameter of a fire hydrant. In an unexpected division of labor, the underground portion of the palmetto seedling folds itself in half, and this is where the saxophone metaphor comes in, although probably it could have been called the elbow—some refer to it as the heel because of its rounded character.

The bottom of the musical saxophone, where it turns, is called the bow. On the palm, this is the rounded base that is pushing down into the soil. Above the bow the young palm is folded with one side sporting roots, and the other, leaves. So when the buried portion of the plant is excavated, it is seen to have two different sides: one side with roots and no shoots, and the other supporting leaves but missing roots. That is the unlikely saxophone. As the tree grows, the root-bearing area becomes ever more massive until the saxophone ends up dwarfed and looking vestigial. Here is Tomlinson's description: "The seedling leaves are reorientated almost 180° and the leaf base develops a bulbous protuberance which assists the axis in forcing its way into the soil. The base of each successive leaf protrudes through a slit in the base of each older leaf. This oblique development results in a characteristic leaf scar on the downward-directed seedling axis. At this stage the 'saxophone' shape of the axis is very clear."[4]

The swollen base of the tree becomes the support for the trunk and the origin for the extensive root system that serves two functions—anchoring the plant so you can't pull it up (even a little one with just a few leaves) and storing reserves so that when you mow over it, or shear it off, it has the resources to sprout up again. This ability to resprout is augmented by two other traits: they seem resistant to many herbicides, and they frustrate string trimmers by shredding into a rat's nest of tough green fibers rather than simply being sheared off.

This phase of a cabbage palm's life can last decades, but it helps explain why they insidiously manage to intrude into landscapes where pines, oaks, and other species are routinely removed. Each time the cabbage palm re-sprouts it is gaining—increasingly able to take advantage of a period of inattention before they "suddenly" appear as a legitimate part of the landscape.

Young cabbage palms sport all fronds angling upward from an origin at ground level, so after small cabbage palms are more than a few feet high they look more like ground-based, fountain-style pyrotechnics that send up a shower of sparks. John Muir describes it as well as anyone: "They stand erect at first, but gradually arch outward as they expand their blades and lengthen their petioles."[5]

Once the trunk emerges, the palm grows more rapidly—a growth spurt of sorts. As a result of all these strategies, some (like John Muir) are led to conclude the palm grows rapidly, but this is analogous to the musician who plays in roadhouses for a quarter century and then is heralded as an overnight sensation. The necessity of amassing an adult-diameter trunk with juvenile leaves makes the cabbage palm uniquely adapted to invad-ing landscapes that humans routinely assault with hand-pulling, mowers, string trimmers, and herbicides, all of which have only modest effects on cabbage palms. As a consequence, cabbage palms are better equipped to invade neglected landscape areas than are pines or oaks. The result is palms popping up in landscape beds, lawns, and adjacent to fixed objects such as tree trunks, fences, and phone poles, where they frequently outlast land-scapers' efforts to defeat them.

5

———•———

One Plant, Many Names

Sometimes the most obvious or common things are the hardest to define or name. Obscure body parts we don't refer to often (philtrum, glabella) tend to have specific narrow meanings, whereas terms in common parlance seem vague. Where does a cheek become a temple or a jaw? Plants, like human body parts and animals, frequently have common vernacular nicknames as well as more formal, technical, scientific names. The prevailing system of scientific naming is a system of binomial (two name) nomenclature pioneered by Swedish botanist Carolus Linnaeus in the second half of the eighteenth century. If there is one thing people know about Linnaeus, it is that he was central in envisioning a system that would allow everyone to know unambiguous names for organisms—potentially disassembling the Tower of Babel that prevented people from knowing precisely what plant or animal was being discussed.

John Kunkel Small was a standout in the pantheon of southeastern botanists, particularly in southeast Florida. He spent lots of time in the field, wrote accessible descriptions of his findings, linked his work to that of other botanists, and was also an early conservationist. In his 1923 article "The Cabbage Tree—Sabal Palmetto," he asserted that Linnaeus knew about cabbage palms because in 1765 Linnaeus wrote a letter comparing seeds of one plant to "the seeds of Palmetto from South Carolina." But, despite the reverence we may hold for Small, he was mistaken. Linnaeus didn't write the letter; he received it from a correspondent, John Ellis.[1]

So what were those seeds? In geographic terms, the most common palm in South Carolina is *Sabal minor,* a small, typically ground-hugging palm. The most visually conspicuous palm in South Carolina is the *Sabal palmetto,* an obvious palm tree, and common along the coast in South Carolina, yet historically rare elsewhere in the state. And saw palmettos, *Serenoa repens,* are also found in the southernmost corner of the state.[2] So not only is it ambiguous what plant Ellis was referring to, but it is not clear that Linnaeus understood the reference since he didn't mention it in his reply.

South Carolinians routinely refer to *Sabal palmetto* as palmettos, apparently with relatively little chance of confusion, because scrub palmetto, *Sabal etonia,* isn't found there, and saw palmetto, *Serenoa repens,* is quite rare. In Florida, in addition to cabbage palms, the term "palmetto" can frequently refer to saw palmettos, *Serenoa repens,* as well as dwarf palmetto, also known as blue stem or swamp palmetto, *Sabal minor,* and scrub palmetto, *Sabal etonia.*

This problem, which I think of as "Ambiguous Palmetto Reference," vexes attempts to understand the palms of the North American Southeast. I can attest that nothing complicates historical research regarding cabbage palms as much as explorers' fairly predictable and casual reliance on the infuriatingly ambiguous term "palmetto." Consequently, dozens of palmetto references are either hopelessly compromised or rely on sketchy contextual assumptions.

Further complicating matters are the following fonts of confusion:

- Swamp cabbage is also a common name for skunk cabbage, *Symplocarpus foetidus,* a stinky wetland plant with a range that doesn't overlap with cabbage palms.
- Cabbage-tree palm is also a common name for *Livistona australis,* an Australian palm that also can be harvested for its edible "heart."

Cultures frequently use multiple common or vernacular names for commonly encountered living things, and, thanks to Linnaeus and others, we now have agreed-upon scientific names. Common names are frequently linked to descriptions the general public can interpret, whereas scientific names travel with technical descriptions relying on specific anatomical and descriptive terms, in the case of plants, terms frequently known only to botanists. In addition to written descriptions, illustrations provide a third means for characterizing a plant or animal. In English, cabbage palms are commonly called palmettos, sabal palms (and sable palms!), swamp cabbage, Carolina palmetto, or cabbage palmetto.

What were cabbage palms called before Europeans arrived? Kanapaha is a locale and the name of both a botanical garden and a plantation in Alachua County, Florida. The Historic Haile Homestead (at Kanapaha Plantation) claims Kanapaha is "Indian for 'small thatched houses.'"[3] The Kanapaha Botanical Gardens website asserts that "Kanapaha" is a Timucuan word for palm house derived from cani (palmetto leaves) and paha (house).[4] Further confusing the situation is the fact that cabbage palms are called "palma cana" in Cuba.[5] So, Kanapaha aside, are there any other cabbage palm-related words or place-names inherited from the Apalachee, Calusa, Tequesta, Timucuans? Hard to say. The arrival of Europeans came close to erasing somewhere between ten and twenty thousand years of accumulated knowledge about the New World. The Spanish seemed especially inclined toward subjugating, converting, and killing natives (deliberately or inadvertently via infectious diseases). Many of the original inhabitants who survived did so marginalized in new unfamiliar locations, new ecosystems without the accumulated local knowledge of more vulnerable elders.

As a result, there may be hundreds of irretrievable observations about cabbage palms alone that are now forever beyond our grasp. Daniel Austin's nine-hundred-page *Florida Ethnobotany* needs only two pages to deal with all three *Sabal* species.[6]

In Florida, original tribes were replaced by Muscogee Creeks, Yamasee, Hitchiti, Oconee, Yuchi, Tamathli, Chiaha, and others and became today's Seminoles and Miccosukees. And despite disease, forced relocation, and three wars, the undefeated Seminoles fared better than many other North American tribes. We know far more about these more recent tribes; for instance, Seminoles are recorded as calling cabbage palms tah-lah-kul-kluk-ko, tala-la-kulke, ta-la-thak-ko, or talaofo.[7] Were these regional variations, mishearings, transcription errors, or did different variants convey nuanced information about the palms in question? Present-day Seminoles may have some insights—another thread to untangle.

So what is the origin of the scientific name *Sabal palmetto?* The *palmetto* species name is fairly easy to trace. European awareness of, or interest in, cabbage palms unfolded slowly over several centuries. The first three significant Spanish explorations of North America approached through southwest Florida from bases in Cuba and the Caribbean. Juan Ponce de León sought "gold and glory of conquest" and, having conquered Puerto Rico, arrived in southwest Florida twice, once in 1513 and again in 1521 with two hundred settlers, animals, plants, and seeds—the basis for a permanent settlement.[8] They were constantly harassed by natives, and when Juan Ponce was wounded, they withdrew to Cuba, where he died from his wound.

In 1528 Pánfilo de Narváez landed with three hundred men, also looking for gold. Only four men survived, including Álvar Núñez Cabeza de Vaca, who reported, "We marched for fifteen days, living on the supplies we had taken with us, without finding anything else to eat but palmettos like those of Andalusia."[9] Several pages later he mentions "low palmetto, like those of Castilla."[10]

Then Hernando de Soto tried his hand. He had helped conquer the Incas and knew how to subjugate natives. On May 30, 1539, twenty-six years after Ponce de León "discovered" Florida, the De Soto expedition landed in Florida, somewhere on the southwest coast with six hundred soldiers, 223 horses, war dogs, and a retinue of pigs.[11] According to the Gentleman from Elvas, the next day they arrived at the town of Ucita, where they found houses of wood "covered with palm leaves."[12] Near present Dade City they found their first cornfields,[13] emphasizing that what had been Spanish conquests elsewhere in the New World might more accurately be termed corn quests in Florida as both de Narváez and de Soto, traveling with hundreds, were forced to keep moving to find sufficient food—maize the natives had been growing extensively since AD 1000.[14] At a location called Cale, "they found palm cabbages in low palm trees like those of Andalusia."[15] So were the Spanish eating swamp cabbage?

The Spanish were familiar with palms (which they called palmas) if not from Cuba, the rest of the Caribbean, and Central America, then from Spain, where there grows a palm more or less intermediate in appearance between North America's saw palmetto and cabbage palm. Like saw palmettos, they have armed petioles, and while typically clumping, some individuals have a single trunk, which seems similar to a cabbage palm. The species is *Chamaerops humilis,* known as European fan palm or Mediterranean dwarf palm.[16] Significantly, the heart, known as "palmito," is consumed in Spain.[17] So the diminutive of "palmas" produces "palmitos," which no doubt accounts for New World "palmettos," both small and large.

With palmetto borrowed from the Spanish, what about *Sabal,* the name of the genus? English-speaking botanical explorers seem to have been more curious about cabbage palms than were the Spanish. English naturalists who contributed to understanding of the cabbage palmetto include John Lawson, Mark Catesby, and Thomas Walter.

In 1700 English naturalist John Lawson started exploring from Charleston and eventually published his observations in 1709, describing palmettos as follows: "The Palmetto-trees, whose Leaves growing only on the Top of the Tree, in the Shape of a Fan, and in a Cluster, like a Cabbage; this Tree in Carolina, when at its utmost Growth, is about forty or fifty Foot in Height, and two Foot through: It's worth mentioning, that the Growth of the Tree is not perceiveable in the Age of any Man, the Experiment having been often try'd in Bermudas, and elsewhere, which shews the slow Growth of this Vegitable, the Wood of it being porous and stringy, like some Canes; the Leaves thereof the Bermudians make Womens Hats, Bokeets, Baskets, and pretty Dressing-boxes, a great deal being transported to Pensilvania, and other Northern Parts of America,(where they do not grow) for the same Manufacture. The People of Carolina make of the Fans of this Tree, Brooms very serviceable, to sweep their Houses withal."[18] This is the first detailed description of a cabbage palm I've encountered, and while it conflated Bermuda's *Sabal bermudana* with the cabbage palm, it emphasizes the slow growth of the species and its utility for crafts.

In 1722 English naturalist Mark Catesby also started exploring the region around Charleston, South Carolina. Catesby was as effusive as the Spaniards were terse, and his first comments regarding their slow growth and their utility were clearly based on *Sabal bermudana* and Lawson before him. In 1731 he described the "Plat Palmetto": "This kind of Palm also grows on all the maritime part of Florida and South Carolina, whose northern limits being in the latitude of 34 North" (which was remarkably accurate). He then went on to state that they diminish in size as one moves north, claiming, inaccurately, that these shorter variants can be found "as far North as New England."[19] In 1767 Catesby published his *Hortus Europae Americanus* wherein he suggested palmetto trees might be grown in England (he was right), mentions that they can be found on Sullivan's Island, and notes the "leaves are used for the walls and covering of houses." That volume also included a colored engraving of "the palmetto-tree of Carolina" with what is likely the first colored depiction of a cabbage palm that purports to show detail, although the base of the trunk is atypically swollen and the petioles puny.[20]

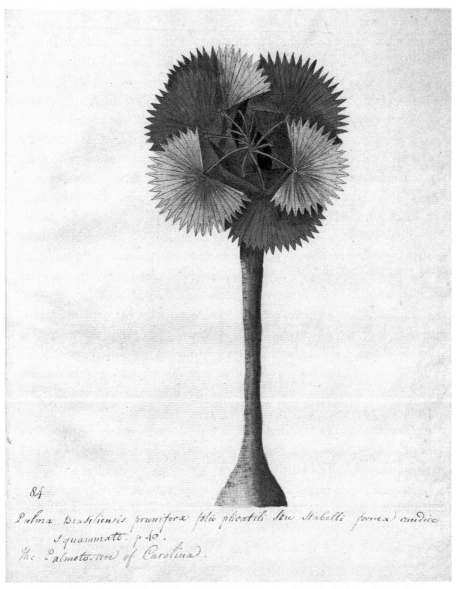

84

Palma Brasiliensis prunifera folio plicatili seu flabelli forma caudice squammato. p 40.

The Palmeto-tree of Carolina.

Mark Catesby's version of a cabbage palm features an atypically swollen trunk base and flat, fanlike fronds. Biodiversity Heritage Library, Missouri Botanical Garden, Peter H. Raven Library.

In 1756 P. Browne described what he called Corypha palmacea assurgens foliis flabelliformibus semipinnatus, petiolis majoribus compressis, using the system known as polynomial (as opposed to binomial) nomenclature. It was a Jamaican palm, what we now call *Sabal maritima.*[21] So Brown was the first to technically describe a *Sabal* palm, just not *Sabal palmetto,* and he wasn't using the Linnaean binomial system.

On December 31, 1765, John Bartram wrote about "the tree palmetto, which the inhabitants call cabbage-tree."[22]

It was German-born William Young who provided another image of a "cabbage palmetto" also in 1767. He was, by his own admission, not a great artist, and he provided his cabbage palmetto with red, not white, flowers.

Catesby, Young, and F. A. Michaux (*Cabbage Tree,* 1819) all created color illustrations of cabbage palms. None is particularly accurate, although by viewing all three in concert, one could conclude these trees have straight brown trunks, no green crownshaft, and generally palmate leaves with unarmed petioles. In their time these illustrations were probably as eagerly awaited as the first pictures from the surface of Mars.

Linnaeus's contemporary, French botanist Michel Adanson, was unimpressed with the Linnaean binomial system of organization and developed his own competing approach. In 1763 he offered the single name "Sabal" for a species of palm from the Carolinas with no explanation of the name.[23] He noted it as "palma caroliniana," with the English annotation: "Swamp palmeto," so it was undoubtedly *Sabal minor.* Dutch scientist Nicolai Josephi Jacquin actually provided the first official Linnaean binomial name of what we now know as a *Sabal* palm, in Latin, in 1776. He called it *Corypha minor. Corypha* is an Old World genus.

Because Adanson's system did not persist, advocates of Linnaean binomial nomenclature could have simply done away with "*Sabal*" since iconoclastic Adanson hadn't followed the taxonomic rules. But in 1804 Frenchman Louis Ben Guersent published a description using *Sabal* as an accepted name for a genus, and he honored Adanson with the formulation *Sabal adansonii.* Consequently Guersent gets credit for establishing the *Sabal* genus name, when he described *Sabal minor.*

British-born American botanist Thomas Walter attempted to survey the plants within fifty miles of his residence in South Carolina, and it is Walter who gets credit for naming the plant we now know as *Sabal palmetto.* Walter called it *Corypha palmetto.* Then, in 1823, Conrad Loddiges, realizing

Sabal palmetto, cabbage palmetto. Watercolor by William Young, 1767.
Held in the Botany Library at the Natural History Museum, London.
The Trustees of the Natural History Museum London.

A. Riché del.ᵗ   Renard sc.

Cabbage Tree.
Chamærops palmetto.

Michaux's *Sabal palmetto* has black fruits, rudimentary bootjacks, and slightly curved fronds—an improvement over earlier images. From *The North American Sylva,* vol. 3.

it wasn't an Old World *Corypha*, went out on a limb and listed *Sabal palmetto* in his nursery catalog. Finally, in 1830 "Schultes & Schultes first validated the new name and provided a description, citing Walter's basionym and giving credit to Loddiges for placing it in Sabal."[24]

As David Fox summarized in his dissertation, "In his monograph on Sabal, Zona (1990) describes the taxonomic 'trip' imposed on *Sabal palmetto*, from when it was likely first described in 1756 as a member of *Corypha*, moved to *Chamaerops* in 1803, confirmed to be a *Sabal* in 1830, uprooted and moved to *Inoides* in 1901, and finally restored to *Sabal* in 1907."[25]

According to Scott Zona, there are currently fifteen species of *Sabal* palms and three times as many names for them that are no longer used.[26]

The reader may feel entitled to a coherent, intriguing, and definitive story about the origin and meaning of the genus name *Sabal*. None exists.

In the absence of any Adanson-provided backstory about the genus *Sabal*, people have become ridiculously creative. So creative and outlandish that I've been encouraged not to repeat them. The list includes anagrams,[27] colors,[28] and a French word for sand.[29] None seem very plausible.

Bottom line: there is no accepted explanation for how Adanson created, or appropriated, *Sabal* as a genus name. Intriguingly, a book on the palms of Cuba, *Las palmas en Cuba,* where cabbage palms are also found, offers (when translated), "The generic name Sabal has its origin in the word with which the Indians of southeastern North America called this palm."[30] Perhaps some indigenous cabbage palm words did persist, hidden in plain view?

Part II

6

———•———

Anomalies

Oddball Palms Get Noticed (and Collected)

Cabbage palms, as a rule, have one canopy, green leaves, a relatively straight, upright trunk of a fairly constant diameter, grow vertically, and have their roots in the ground. So when, of the millions of cabbage palms, any are found that diverge from this norm, some observant people take notice.

It is not entirely clear why we, as humans, get so caught up in collecting, displaying, and generally obsessing about anomalies. Sometimes the anomalous is simply the last-of-its-kind rarity, whether it was Lonesome George the Galapagos tortoise or a Honus Wagner baseball card. Sometimes these obsessions can reflect an unhealthy fascination with the abnormal—two-headed calves, a contorted cactus, or the sideshow Lobster Boy. We are drawn to the anomalous simply because it is unexpected, it breaks the normal rule. From Victorian curiosities to Ripley's stock in trade, we seem to be on the lookout for whatever is different.

Unanticipated color variants fascinate us. Purple carrots, pink grapefruit, and striped beets awaken our plates, if not our palates. Plant breeders are always on the lookout for a new flower or foliage color, whether a new rose or orchid, or leaf colors or shapes that deviate from the norm. And an absence of normal pigmentation, albinism, has long fixed our attention, whether it is Melville's white whale or more contemporary examples such as Snowflake the Gorilla or any number of leucistic alligators on display in zoos and aquariums.

All plants that get their energy from the sun need chlorophyll, and, consequently, there are no photosynthesizing albino plants. But plants lacking chlorophyll in parts of their leaves do exist, and some plant collectors crave such variegated varieties. While it is not hard to find cabbage palms that exhibit a certain dull-yellow paleness near the costa, truly variegated cabbage palms are rare. In these individuals the depigmented areas are much more dramatic and extend through the leaf segments.

In 1968 a description of a Fort Myers garden tour mentioned "a tall variegated cabbage palm with its yellow and green striped fan-shaped leaves" at a home on Vesper Drive.[1] In 1994 cabbage palm researcher Kyle Brown visited the Riverwood Golf Course at El Jobean, Florida, because the manager wanted to show Kyle an unusual variegated palm on the golf course. Brown learned that palm collector Ted Langley had previously gotten seeds from another variegated palm in the area and that Langley was actually growing variegated seedlings. Brown effected a palm exchange and started growing his own variegated cabbage palm.

Brown notes that there are two types of variegated palms—a striated variant with yellow stripes running along the leaf segments (apparently more common on the east coast) and individuals with bold patches of glowing paleness centered at the costa, where the segments originate—these are the type seen in the Sarasota/Bradenton area. It is probably safe to say that several dozen of these sunburst palms have been found in the Myakka and Braden River watersheds, but they don't always fare well. First, since many of the sightings are through windshields, these examples fall victim to road widening and development. Those that survive progress are at risk from naïve collectors unaware of the challenges of relocating young cabbage palms. A young one in Myakka River State Park was nearly killed by mowing of the shoulder of the park drive. In addition, I've been told that some anxious or uninformed landscapers conclude the yellowed palms are diseased and dispatch them. And, since they don't have as much photosynthetic surface as a normal palm, they may be a little less vigorous. In 2007, the Selby Botanical Gardens acquired one with some fanfare (a newspaper article suggested it might be worth twenty thousand dollars),[2] but it did not survive.

I spotted one growing along Highway 72 in Sarasota County, and I was told the discoloration was likely the result of herbicide spraying or a disease. But it persisted for years, and when road widening seemed imminent I persuaded the county government to relocate it to the eastern terminus

The author poses with dramatically variegated palms on Seventeenth Street, Sarasota, Florida. Photo by Julie Morris.

of Valencia Road in Venice. The most dramatic examples may be viewed on Seventeenth Street in Sarasota. Owned by Ace Holland, these three palms (and now a bevy of offsprings) are irrefutably variegated, and Ace has started collecting the seed for a grower who is trying to develop them commercially. I've seen seedlings from this operation for sale online for five hundred dollars. As more palm promoters germinate seeds from reproducing variegated palms the price for seedlings should come down.

In the meantime, palm collectors looking for oddities have contented themselves with another leaf variation that has entered the commercial trade. In 1998, Robert Riefer spotted an odd palm in an interstate highway interchange in Lee County. It was clearly a cabbage palm, but it looked as though the leaf segments never separated in the usual way. He subsequently discovered others, obtained seeds, and found that about two-thirds of the seedlings possessed the same anomaly—slightly contorted leaf segments that fail to open fully.[3] He named this variant for his wife, 'Lisa.' Whatever the nature of this oddity, it is not unique; other plants with the same affliction have been found growing wild in Sarasota, and enthusiasts and arboretums now have affordable Lisas in their collections.

At the other extreme are the cabbage palms (or *perhaps the* palm?) with unfused leaf segments. I don't know how many people would drive 290 miles to look at an oddball palm, but I did. I was going in search of a "split-leaf" cabbage palm, a palm whose leaf segments were not attached to each other. Instead of the palm leaf segments being fused for much of their length, each segment leaves the costa as a long, green, folded streamer, which would be comparable to your fingers extending to your wrist, with no palm.

I found myself in a confusing rural subdivision and eventually realized I should be looking for a palm-oriented front yard. That led me to Lucas and Jasmin. I grabbed my walking stick, and we hopped in their car and drove for about ten minutes to a nondescript road shoulder. We piled out, crossed a shallow ditch, and plunged into dense pine woods. We stumbled around in vines and branches until we found the palm with its ragged leaves. The ground was littered with thin, delicate bootjacks, and the shredded fronds looked like they had been through a hurricane, even though it was in dense woods. I wanted a frond or two, but they were just out of reach of my walking stick. Lucas was young and had some FSU circus arts training, so Jasmin climbed on his shoulders and, using my walking stick as a crook, collected one dry and one green leaf.

One of the more conspicuous cabbage palm anomalies is the polycephalic palm, the one with two or more "heads." There are many species of palms with multiple trunks, but palms with forked or branched stems/trunks are rare. The doum or gingerbread palm, *Hyphaene thebaica,* commonly develops multiple heads, creating a tree worthy of Dr. Seuss. You can see one near the bookstore at the Florida Institute of Technology or several at Fairchild Tropical Botanic Garden. Two-headed cabbage palms are novelties sought by collectors, and, since mature cabbage palms typically transplant well, when discovered in the wild they are frequently dug up and brought into cultivation.

Famed botanist John Kunkel Small spent thirty-seven years exploring Florida at the beginning of the twentieth century and was intrigued by polycephalic palms, which he referred to as being fasciated. Fasciation usually refers to a habit of abnormal plant growth where what is normally a single point of growth appears as a flattened or fused growth along a line. Small took a picture he called *Palm with Split Trunk* in a pine prairie in January 1927.[4] A different Small photo, *Palm with Split (Two Trunks),* is undated.[5] Small's photographic fascination with fasciation may have helped popularize multiheaded cabbage palms as coveted collectors' items. In 1965 the *Fort Myers News Press* ran a photo of a similar two-headed palm discovered by a father and son who had been hunting in the Fakahatchee Strand.[6]

In the late 1950s Julius and Leonard Rosen's Gulf American Land Corporation paid less than four hundred dollars per acre for some land known as Redfish Point, north of Fort Myers, on the north side of the Caloosahatchee River. That became the nucleus of what is now known as Cape Coral, a speculative lot subdivision (now city) eventually covering 114 square miles.[7] Part of the bait to get visitors to become buyers was Cape Coral Gardens, a Whitman's-sampler attraction containing, among other things, a cast stone version of the flag-raising on Iwo Jima, replicas of both Mount Rushmore and Michelangelo's *Pietà,* a "Porpoise Paradise" pool and show, forty thousand rose bushes, a carillon, the Waltzing Waters synchronized fountain display, an animal sanctuary, and a two-headed cabbage palm.[8] And today, if you drive down Rose Garden Road, about a half mile past the intersection with El Dorado, you'll see the two-headed palm, which apparently is all that is left there of Cape Coral Gardens (the Waltzing Waters attraction relocated). The site was too valuable as waterfront residential real estate, and, according to 2017 Zillow prices, the nine waterfront homes closest to the remaining forked palm are now worth more than what Rosen paid for the initial 1,724 acres.

The first double-header I ever saw was growing wild in a palm/oak hammock on the west side of the Myakka River. Farther south another two-headed cabbage palm overhung the banks of the Myakka for many years

John Kunkel Small's *Palm with Split Trunk*, January 1927. Courtesy of the State Archives of Florida.

and became the subject of a poem by conservationist Mary Jelks.[9] It's gone now, but its presence, combined with one on a nearby ranch and another farther south, led one author to speculate that the two-headed palms were created deliberately by Native Americans as some sort of signposts[10]—an appealing, if unlikely, narrative.

Julian Dimock photographed a tall two-header on Marco Island in 1906.[11] Two Koreshan women appear in an undated black-and-white photo in front of a "split palm."[12] If you want to see one in the wild, there's a trail at the Alafia River Corridor Nature Preserve with one,[13] but, once discovered, many get relocated to more urban settings, where they don't always fare well. An old postcard shows one at the College Arms Hotel in Deland.[14] Although it could have been transplanted, it was lost during redevelopment. The one that had been at the John D. MacArthur Country Club in Palm Beach is also gone.[15] There was one in the Florida Panhandle at Indian Pass, before you got to the boat ramp, but it did not survive Hurricane Michael. There is one at the Palm Bay, Florida, entrance sign,[16] but the comparable one donated by Robert Riefer, of 'Lisa' fame, for the entrance of Lehigh Acres in 2012 didn't make it—by 2017 it was gone.[17] A bank in Palmetto, Florida, featured one for many years, and it's gone but was figuratively replaced by another growing at a Goodwill store also in Palmetto, which was subsequently moved to city hall. There were two on the University of Florida campus in Gainesville, but the one by the infirmary is gone, survived by the two-headed "bicentennial" cabbage palm west of the student union. A precautionary metal rod links the two trunks. The Sanibel fire station has a great one, which, like the one in Palmetto, split close to the ground.

The most reliable places to find twin cabbage palms are botanical gardens. A Steinhatchee father-and-son team actually found two, one in 1981 and one in 1990, and delivered them both to Kanapaha Botanical Gardens in Gainesville. Two-headed cabbage palms may also be seen at Cypress Gardens,[18] Ravine Gardens in Palatka,[19] and McKee Botanical Gardens.[20] These are special, rare trees, but if double-headers occur only once in a million, there should be 108 double-headers in Florida alone.[21]

The relative ubiquity of two-headers inevitably begs the question of three-headers, and there was one on Volusia Avenue in Daytona, but that seems to persist only as a postcard. There was another in Deland, and John Kunkel Small photographed one in Daytona Beach.[22] The only contemporary three-header I'm aware of is at Heathcote Botanical Gardens in Fort Pierce, Florida. According to a sign in the Gardens, it was moved by oxcart from Okeechobee to Fort Pierce in the early 1900s. It has been moved a total of five times and is estimated to be more than 140 years old.[23] The most famous three-header may have been the one at Kanapaha Gardens, which was also found by that Steinhatchee duo, Bobby Futch Jr. and his father, Bobby Sr. They found it near Perry in 1994. Curiously, the tree had once been a four-header, but by the time the Futches had gotten to it, it was down to three. They successfully moved it to Kanapaha, but it lost two more heads in a storm.

Rural Dixie County is on Florida's Gulf coast. It has remained rural because it lacks sandy beaches. The *Dixie County Advocate* has been published every week since July 1, 1921, and the masthead makes only two claims: "The Only Newspaper & Official Organ of Dixie County" and "Home of the World's Only 4-Headed Swamp Cabbage." Neither TripAdvisor nor Yelp seemed to have caught on to this second distinction, so I wrote to the Chamber of Commerce for assistance in locating this rare tree. Driving through Cross City we failed to find the Chamber of Commerce offices, but I spotted the *Advocate.* So we stopped and I queried the *Advocate* staff while my wife, Julie, inquired at the courthouse. The two women staffing the paper weren't sure about the location but recommended Angel at the nearby Dairy Queen, who, they assured me, knew right where it's at. Meanwhile Julie tried the property appraiser's office and was told the four-header was out in a swamp. A voice from another room suggested directions to another, more accessible palm (could Dixie County be the home of the only two?), and, after learning Angel's DQ shift had ended at four o'clock, we made our way to SE Seventy-First Ave. (aka Rudolph Parrot Rd.), turned east, and almost immediately found the world's only four-headed swamp cabbage on the front lawn of the Cannons, who had already let the Chamber of Commerce know we were welcome to take photos. I thought it looked like a lyre since a wooden brace resembled a crossbar and the two outermost trunks appeared to bend outward. A Cannon in-law related that Mr. Cannon had discovered it while hunting turkeys. He had stopped by the palm to try a turkey call, and only after he left did he glance back and notice the four trunks. Not only that, but there was a two-header next door.

Unfortunately for everyone who thought their four-header was unique, I kept discovering more four-headers. Luckily, it wasn't until after our visit to Dixie County that I discovered Kyle Brown had photographed another four-header. Could Brown have photographed the Dixie County specimen? No, because multiple-headed palms seem to occur in two different forms. The Dixie County tree has all four trunks in one plane or axis, sort of like a deficient menorah. The palm Kyle Brown photographed had trunks in multiple planes. Apparently there had been one at Silver Springs that had forked once, and then each fork diverged a second time.[24] There's also another menorah-style four-header in southern Hillsborough County. The owners claim to have been offered nineteen thousand dollars for it,[25] and I suspect this is the one that is said to have appeared in an edition of *Ripley's Believe It or Not*.

Now you are probably asking yourself, has there ever been a five-header somewhere? Yes, and guess who photographed it in 1917 in Lee County? John Kunkel Small. Just another of his fasciated palms.[26] There's another even stranger one in Venice, Florida, that looks like a clump of cabbage palms planted together, but, upon inspection, this clump has six trunks. The one on the south has two heads, the next has one, the next has three, but one of those split again for a total of four, then a three-header and then a one—for a total of eleven canopies.

I found a picture of a ten-header online, but, when I queried the photographer, I was dumbstruck to learn he had forgotten where he had taken the image. There was a fence and an entrance gate, and I assumed it was a private ranch I would be unlikely to stumble upon. Several years later I was online looking for images of Tom Gaskin Jr.'s palm log home in Palmdale, Florida, and came across what looked like the same ten-headed palm. That took me to the "Only sabal palm like it in Florida at the Park." A park? What park? Further online sleuthing revealed it was located at the Sabal Palm RV Resort. So then it was off to Google Earth to search the Palmdale area for an RV park. I found a small RV park, switched to street view, and there it was at the entrance. The top-heavy palm had blown over in Hurricane Irma and suffered in the process. When I got there about a half year later, the tree had been righted but did not look good. There were five dead canopies, several that had been cut off (and presumed dead), two "sockets" where trunks had detached leaving concave root masses, and nine live, but challenged, canopies. It's possible that had I gotten there a year earlier, I might have seen a nineteen-headed palm, although the situation was confused by two separate trunks emerging from the ground. It appeared to me the tree had been righted and left to fend for itself. I subsequently called the front desk and advised them to water it persistently, but, after hanging up, I had no confidence my advice was getting through.

As to how and why some palms develop two or more heads, there is agreement that something happens to the single bud that causes it to split into two growing tips, but theories vary regarding the natural causes of polycephaly. Some suspect lightning or fire, others, insect damage or cold weather. But if it were cold, the prevalence should increase as one moves north. And while Tybee Island, Georgia, has "some of the cutest cottages on the planet and the two-headed palm tree (5th Avenue near Chatham Avenue),"[27] and there's one on Fripp Island, South Carolina,[28] two-headed palms don't seem especially plentiful north of Florida. In 1994 landscaper Dean Vanderbleek hypothesized that "someone desiring a two-headed palm tree could create one by smashing the growth buds of enough palms until one survived the blow and formed multiple heads."[29] I lost track of a gentleman who was doing just that—taking an axe to cabbage palms in the hope of damaging the growing tip in just the right way. I was skeptical but later learned of a company on Pine Island, Florida, that claimed to be creating two-headed palms: "In nature, a palm may become damaged and then grow to form two heads. Growers mimic this damage by splitting the tree and then caring for each head as if it was a separate tree."[30]

That called for a road trip, so I enlisted my good friend Rob Patten, who has occasion to buy cabbage palms wholesale, to venture to Pine Island, where Troy Randall spent a couple of hours driving us around the expansive PalmCo wholesale palm and bamboo nursery. He drove us past impressive windrows of dead coconut palms, pushed over because they could not be sold—liability concerns about the falling fruit have severely depressed the market. Troy told us about a variegated coconut he naively sold when he was getting started. That led to Troy lamenting about a gorgeous, profusely variegated cabbage palm he found at another nursery he had worked at previously. The palm was small, and he asked a field hand to root-prune the tree. His request was misinterpreted, and the worker dug the palm up, dooming it because of its small size. According to authors of

Ornamental Palm Horticulture, Sabal palmettos "are virtually impossible to transplant until they have a well-developed trunk." Palms with less than 6 feet of trunk "have very high mortality rates when transplanted," whereas those with 10-foot trunks "usually transplant well."[31]

Consequently, I've heard 8 feet of trunk cited as a minimum for success, but occasionally naïve gardeners, unaware of these definitive rules of thumb, plunge forward and successfully transplant smaller trees. The paradox of bigger trees being easier to transplant successfully than small ones seems to lie in the fact that the tree has to grow a new set of roots and a large (10-foot) trunk can keep the leaves hydrated for one hundred days, whereas a 3-foot trunk can't manage a week.[32]

I asked Troy about the special two-headed palms PalmCo was advertising online. At first he thought I was talking about double palms—two trees grown in close proximity that are transplanted together as a pair. Once I conveyed that I was talking about forked trees with a single trunk, he drew a blank, suggesting that unusual trees could be special-ordered from folks who looked for them in the wild. The talk of multiheaded palms brought back a memory for Troy—his father had told him of a thirteen-headed sabal palm near the old fiber factory in Gulf Hammock. Is it still there? Is there a fourteen-header somewhere?

The most common divergence from the cabbage palm norm involves trees that possess curved trunks and that can increase their value.

In Florida, at least, plant nurseries have standards or grades for plants they sell: typically, Florida Fancy, Florida #1, and Florida #2. These standards were originally adopted by the legislature in 1955. All trees sold in Florida are to be graded according to provisions of the act, and those not meeting the standards of the top three categories are deemed culls and are not to be sold. For instance, an oak with a trunk dogleg that deviates more than one times the trunk diameter cannot be sold as Florida Fancy or Florida #1. The clear implication is that quality trees are expected to have straight trunks. But a loopy cabbage palm can be graded Florida Fancy and, perhaps perversely, curved cabbage palms are not only allowed but fetch higher prices than perfectly straight individuals.

There are two ways to obtain palms "with character" or "supercurve." The first is to search pastures and hammocks for oddball palms. Rob and Troy agreed that task could not be left to minimum-wage workers but needed someone with a clear understanding of how dramatic a palm needed to be. The second approach was to transplant straight palms into the nursery and plant them at 45-degree angles with the use of braces. Eventually the palms turn upward to create supercurved specimens that sell for much more.

Selling an oak with two leaders would be prohibited, but in 2008, Land's Palm Trees was even selling a double-header. So anomalous cabbage palms are judged by a different, some might say double, standard that accommodates individuals' desires to have unusual palms, with character.

Why do some cabbage palms curve? In many cases, the explanation for not growing ramrod straight is obvious—like a plant growing in a barber shop window, the tree may simply grow toward the light, a positive phototropism. In other situations, the tree tipped over, and the bud made a slow turn toward growing vertically. Cabbage palms tip for two primary reasons. The first is because the trunk and canopy are blown over while enough of the roots remain for the tree to persist.

There's a fascinating grove of tipped-over palms in Stuart, Florida, in Flagler Park, which is an exposed point on the St. Lucie River just a few miles from the Atlantic. These trees, and there are at least a couple of dozen, were undoubtedly pushed over by a tropical storm or hurricane. I thought it would be easy to pinpoint when (and then calculate how long it takes a palmetto to right itself), but inquiry with a group of old-timers at the nearby Stuart Heritage Museum revealed two things: first, that people tend to remember (and photograph) blown-over structures more than blown-over trees, and, second, that Stuart has been subjected to so many strong wind events that it is difficult to determine which was responsible.

The second reason palms tip over is because the roots are eroded out from under the tree, as happens along rivers. Florida is flat enough that the erosive power of high-runoff events is moderated, the result being that rivers meander somewhat more languorously, which enables eroding palms to fall in slow motion. As the waterward side of the root mass is exposed, the partially unsupported tree tips toward the river. And in many cases, when the erosion is modest, instead of simply falling in, the remaining roots continue to anchor and nourish the tree so that the bud starts turning upward (a negative geotropism). The result on several rivers is palm trees growing nearly parallel to the surface of the river. This phenomenon, more than any other, demonstrates the extensive root systems of cabbage palms because the leverage created by maintaining a tree at an angle is significant and could not be accomplished with a meager root system. Consider: it doesn't take much strength to balance a broom vertically upside down on the palm of your hand. But it is not easy to hold the same broom horizontally from the top end, despite the fact that it weighs the same in each case. Even with its waterward root mass gone, the cabbage palm hangs on and appears to levitate above the river, held from collapsing by long tension cables (roots) extending dozens of feet deep in the riverbank. These slowly falling palms are iconic and typify rivers such as the Myakka. During higher flow these parallel palm trunks can be just a foot or two above the surface of the water, creating "sweepers" that threaten to allow canoes or kayaks underneath while scraping off the protruding paddlers. In flood stage these parallel palms can become submerged, and the force of the water can push the trunk down until it rebounds and springs back up, creating an eerie, rhythmic bobbing action known to paddlers as "sawyers," apparently reminiscent of the tiller (upper) man's motions in early pit saws.[33] Winslow Homer's paintings captured both leaning and horizontal palms in Florida.

In addition to falling palms making contact with the ground or a river, some palms succumbing to wind or gravity inevitably make contact with other trees. In the most extreme cases, the palms are caught by neighboring trees. One tree in Myakka River State Park fell in the crotch of a laurel oak. The oak grew around the palm, and now the palm enters one side of the oak and emerges on the other. But despite shade and winds and erosion and gravity that explain why some palms bob and weave, there are others that wander about with no obvious explanation.

There are other trunk oddities, eroded trunks, dramatic necking down, and roots perversely emerging far above the soil level. Small photographed a living palm with most of the trunk missing.[34] The necking down is called penciling (since it can resemble a pencil point), and it occurs when the tree is in decline. It creates an hourglass constriction if the palm survives and rebounds. This is a common phenomenon when cabbage palms are transplanted. If you look at transplanted palms in urban areas, you'll likely see constrictions that reflect the stress of transplantation and subsequent recovery. Sometimes the new trunk is more robust, suggesting soil conditions more to the palm's liking. Researcher David Fox believes the transplant recovery constriction can function as "a 'date stamp' of when the

transplanting occurred. If one knows the year when the palm was transplanted, then measurements can be made to estimate the annual growth
rate since transplantation."[35]

Roots that don't originate from other roots are called adventitious
roots. And while palms have rootlets, all main roots originate from the
stem/trunk and, consequently are considered adventitious. The area of the
trunk or stem where they originate is known as the RIZ, the Root Ini-
tiation Zone. As one might expect, the RIZ is typically at or below the
ground surface, but sometimes a RIZ develops higher up on the trunk.
These higher-than-usual roots are in a drier environment so their growth is
arrested, and they typically don't make their way to soil. It is believed that
these "misplaced" roots could become active if they come into contact with
moister conditions.[36]

Unlike unusual dicot trees, which can be very difficult and expensive to
relocate, anomalous cabbage palms lend themselves to collecting because
they can be relocated relatively easily. So, aside from private collections,
many of these anomalies can be found in botanic gardens, and some, such
as the 'Lisa' and variegated palms, may be purchased as seedlings.

So what do all these oddball, distorted cabbage palms mean? In addition to satisfying our desire for novelty, anomalous palms help us better
understand normal palms. Growing variegated palms from seed helps convince us that variegated palms are not victims of herbicides, as some once
thought. Palms with four heads and palms with two-thirds of their trunks
missing suggest typical trunks are "overdesigned" by several factors, which
accounts for the outstanding wind resistance of normal cabbage palms.
Palms eroding along riverbanks help us understand and appreciate the extent of the root systems. And penciling after transplanting confirms that
relocating a palm, while workable, is a debilitating shock. Sometimes exceptions do prove the rule.

7

———— • ————

Diseases Stalk the Cabbage Palm

Once you take an interest in something, you start noticing it more frequently. This phenomenon, which psychologists call the frequency illusion or Baader-Meinhof Phenomenon, is natural (you may be noticing cabbage palms more now that you're reading this book). Prolonged heightened awareness combined with more nuanced pattern recognition can lead to perceptual expertise, the ability to discern what you're looking for more readily. Humans can develop very sophisticated perceptual expertise. For better or worse, herpetologists see snakes you would have walked right by. Sailors note changes in wind direction and intensity that wouldn't register with the rest of us. And mycologists can spot fungi where most people don't. But mycologists and laypeople alike typically see only the fungal spore-producing bodies (such as mushrooms) that enable the next generation, which would be like finding grapefruits lying on the ground with no grapefruit tree in sight.

And, as the mush in mushroom suggests, many of the fruiting bodies are short-lived, transitory phenomena that decompose or are eaten shortly after releasing their spores. Some actually self-dissolve in a process known as deliquescence. So for most fungi we really only notice them briefly when they are in reproductive mode. Otherwise fungi tend to consist of an inconspicuous webwork called mycelium that permeates or covers whatever the fungus is interacting with. This inconspicuousness is what makes their relationship with orchids and hand ferns so complicated, infuriating, and marvelous.

Consequently, some of the best-known fungi are those that possess a persistent, woody, fruiting body (basidiocarp) that projects horizontally from a tree trunk, and these are called shelf, or bracket, fungi. These are polypore fungi that produce their spores from circular pores rather than from the curtainy gills we associate with most supermarket mushrooms.

In addition to the portabellas, agaricus, and the ever-increasing list of specialty edible fungi available, other species are extremely beneficial. Saprophytic fungi break down dead wood to recycle nutrients. Other fungi function as crucial metabolic support systems for many species of trees and other plants that have developed mutually beneficial relationships with specific mycorrhizal fungi that provide nutrients and increased water supply in exchange for sugars.[1]

One fungus that attacks pineapples, sugarcane, and bananas in addition to palms is *Thielaviopsis paradoxa,* commonly, but not too commonly, known as *Thielaviopsis* trunk rot of palm.[2] This fungus rots the softer, non-lignified, portions of the palm, which almost always affects the upper third of the tree. Then the weakened trunk either collapses or, on taller, woodier trees, the canopy simply falls off.[3]

The most perplexing and most reviled fungus attacking cabbage palms is *Ganoderma zonatum,* a parasitic fungus that fatally affects many, and possibly all, species of palms. The mycelia permeate the palm and produce enzymes that degrade lignin and cellulose, killing the palm in the process. The common, if borderline vulgar, name of *Ganoderma zonatum* is butt rot, which refers to its habit of afflicting the bottom 5 or 6 feet of the trunk, which is the most lignified. Unlike *Thielaviopsis, Ganoderma* attacks lignin but does not make the trunk soft.[4]

Ganoderma's diagnostic fruiting body is a bracket or shelf fungus (commonly called a conk) that eventually grows out from the trunk. The scientific name refers to the shiny (gano) upper surface (derma) of the conk, which possesses distinctive zones (zonatum) or banding. Butt rot kills its host and can't really be diagnosed until the spore-producing conk appears, and, perversely, the conk doesn't always appear prior to death, so by the time a conk appears to produce millions of basidiospores, the fungus has already done its damage and the palm is doomed. Conked out. The earliest symptoms that precede the conk are wilting and browning of all the leaves except for the vertical spear leaf.

Replanting another palm in that location is not recommended because palms replanted in the same location frequently also succumb.[5] The conditions that favor butt rot are unknown, and given its modus operandi, perhaps the most perplexing question is why it doesn't infect or affect every palm in a given locale.

Dr. Bruce Holst, the director of botany at the Marie Selby Botanical Gardens, was deep in a cabbage palm hammock on a botanical reconnaissance in Sarasota County, Florida, when he and his crew came upon a clearing in the making—a sunny patch in an otherwise apparently healthy oak/palm hammock where the palms were dying. Several had already fallen, some were standing dead, and others were clearly in decline. And though several of the trunks sported conks, Bruce wanted to rule out another poorly understood disease that can kill cabbage palms. Bruce had invited me along, and we took a trunk sample from a stricken palm. Several weeks later the results came back: negative. So was *Ganoderma* butt rot alone responsible for the all the death and decline? Or was the butt rot a secondary infection of sorts, opportunistically preying on palms weakened by a lightning strike or some other undetected affliction? Were the conks from *Ganoderma zonatum* or some other species of *Ganoderma*? Would the death affect more trees and continue to expand the clearing? The Selby team resolved to revisit the site in an effort to answer some of those questions.

With no means of early diagnosis, no controls, and no cure, *Ganoderma* butt rot is a true wild card in the world of palm diseases. But for all its scary attributes, *Ganoderma* is a known quantity. Like termites, and lightning strikes, *Ganoderma* has been around since whenever and hasn't decimated cabbage palm forests. If *Ganoderma* was the biggest threat to *Sabal palmetto,* there might be more of an effort to study and manage butt rot. But just a few years into the new century, Florida plant pathologists diagnosed a puzzling new disease caused by a phytoplasma.

Phytoplasmas are a type of bacteria (mollicutes) that lack a cell wall. They're devilishly simple—comparable in ways to viruses. They can't survive independent of a host, either a host plant or an insect vector, so they can't be cultured in a lab.[6] While the symptoms of the dying palms are dramatic, a positive ID can only be obtained by sophisticated molecular analyses that have only recently become automated and affordable.

The most famous palm-killing phytoplasma is Lethal Yellowing, which is best known for decimating coconut palms. First detected in Jamaica in the 1800s, it got to Key West in the 1950s and became common in south Florida in the 1970s. Plant breeders have been trying to breed a coconut resistant to Lethal Yellowing for more than a half century, and despite apparent success with Malayan Dwarf or hybrid 'Maypan' (Malayan Dwarf × Panama Tall) cultivars, a truly resistant variety has yet to be identified.[7]

In 1980 another palm-killing phytoplasma was identified in Texas. It had been observed and identified in a declining *Phoenix* palm, so it was given the descriptive, but clunky, name Texas Phoenix Palm Decline, or TPPD. In 2006, TPPD was identified in Florida. It was genetically identical to the phytoplasma found in Texas that had been affecting *Phoenix* palms, so the Texas name stuck. A major outbreak in cabbage palms in Simmons Park on the shores of Tampa Bay alarmed palm lovers, researchers, and growers. Environmentalists were sobered by the specter of TPPD invading native habitats and preserves, where it could completely change the structure of several native plant communities. This was the disease Bruce Holst feared he had spotted in Sarasota County. Nursery growers became alarmed when they learned that TPPD affected not only cabbage and queen palms but also several species of high-end ornamental *Phoenix* palms.[8] This was big news, because while cabbage palms are relatively cheap and not grown from seed in nurseries, Canary Island date palms and sylvester palms are slow-growing, specimen palms and any nursery growing them was now in danger of losing impressive trees worth thousands.

I wanted to learn more about this disease that could, theoretically at least, devastate both the state's natural areas and tree nursery businesses, so in 2017 I signed up for a daylong "summit" hosted by Dr. Brian Bahder in Davie, Florida, at the IFAS Fort Lauderdale Research and Education Center.

It turns out that TPPD is a strange affliction. It does not spread like a deadly wave through a population of palms. The palms that succumb seem scattered through both space and time. While many die, others are spared, only to contract the disease later. Or not. One of the next places the disease was diagnosed in Florida was, ironically, the town of Palmetto on the north side of the Manatee River, south of Tampa. A former citrus grove north of town is bisected by Highway 41, and it's now a cow pasture with both mature and young cabbage palms. I've been driving through there with some regularity for more than a decade, and the view is always the same: some dying palms that were previously green, others with collapsed dead canopies, some just standing trunks, and others nearby, still green and seemingly unaffected. The owner or manager has pushed a number of dead palms into a pile. Will the remaining palms eventually falter and die, or does the genetic diversity of the remaining palms confer some advantage to certain individuals? Or maybe the disease simply hasn't reached the remaining palms. No one seems to know.

If TPPD was more virulent and killed every palm in a concentric outward-propagating wave, it might be relatively easy to secure funds to study the disease. But legislators who control the state's budget are unlikely to be moved if a few palms die in a cow pasture, even if it is the state tree. That's where the clout of nursery growers might kick in. As of 2018, afflicted palms had been documented in twenty-eight of Florida's sixty-seven counties and were no doubt present in others. Dr. Bahder assumes it is now in Georgia. Ominously, he found a TPPD-infected palm in the Everglades at a rest stop along Alligator Alley. That sort of news motivated leaders of the Florida Native Plant Society to get involved in learning more about TPPD.

In trying to build awareness of the threat posed by a disease, it helps if the name is something people can relate to and remember. Zika is a lot easier to remember and discuss than something like Fibrodysplasia Ossificans Progressiva. And for diseases with complicated compound names, such as Severe Acute Respiratory Syndrome, a pronounceable acronym (SARS) is helpful. So when a fatal disease threatens Florida's state tree, it is not optimal to sound the alarm about TPPD.

And for a threat facing Florida, burdening the disease with the appellations "Texas" and "Phoenix" doesn't suggest any urgent geographic relevancy. I even heard a rumor that Texas legislators were disinclined to fund research on any malady that included both "Texas" and "Decline" in the same name. Finally, referring to an inevitably fatal malady with the euphemism "Decline" fails to convey the appropriately serious implications of contracting the disease.

So palm researchers concerned about the threat posed by Texas Phoenix Palm Decline sought a makeover for the challenging name. Florida palm biologist Ericka Helmick suggested Lethal Bronzing, which sounds like it might be the result of sunscreen failure but is playing off the somewhat

Plant pathologist Dr. Brian Bahder with a palm dying from Lethal Bronzing in the background. Photo by the author.

familiar Lethal Yellowing phytoplasma and the fact that the afflicted cabbage palm fronds typically adopt a strange reddish-brown color not normally seen on healthy cabbage palm leaves.

Lethal Bronzing proceeds through several stages as the dying palm shuts down. On mature trees with fruit, the first symptom is sudden and premature fruit drop, followed by the death of any flowers. Of course, these symptoms are evident only on mature plants that have either flowers or fruit. The next symptom is discoloration of the living leaves, starting at the leaf tips of the older leaves. This is where the bronzing phenomenon comes in, although the colors can range from reddish-brown to dark brown or gray.[9] Leaf death proceeds upward, moving from the older to younger leaves. Eventually the spear leaf, the newest upright leaf, fails. Palm pathologists Nigel Harrison and Monica Elliot were quite clear about the implications: "Death of the spear leaf indicates the apical meristem (bud) has died."[10] With only one bud, the palm is doomed.

Of course, there are several ways a palm can die, so a definitive diagnosis of Lethal Bronzing depends on taking a sample from the afflicted tree. Getting the sample involves flame-sterilizing a drill bit and extracting some of the central cylinder trunk tissue without contaminating it. It has to be kept cool and sent by overnight courier to one of two Florida facilities equipped for testing. In 2020 the cost to test one sample was seventy-five dollars. The elaborate procedure and cost reduce the likelihood that suspected palms will be diagnosed, and getting a negative result does not mean the palm is free from Lethal Bronzing—it could be at the earliest stages. Yet testing the palms for Lethal Bronzing is straightforward when compared to looking for it in insect vectors—that involves cutting out the salivary glands of insects smaller than Tic-Tacs™.

In order to manage or control a plant disease, one has to learn how it is spread and how it enters the plant. I went back in 2018 to see what Dr. Bahder had learned. In 2017 he wasn't sure what the insect vector was, although he was confident the disease is spread by an insect that bites into the living palm tissues, which narrowed it down to one of the eighty thousand or so true bugs (Hemiptera). The true bugs include suborders of cicadas, aphids, shield bugs, leafhoppers, and planthoppers. Attack through the leaves appeared to be the point of entry. By 2018 he was quite certain the insect vector was *Haplaxius crudus,* although he had not yet conducted the definitive research. The common name is American palm cixiid, a planthopper no more than one-fifth of an inch long.

Unfortunately, there seem to be three separate ways the disease could move to new locations. Most likely: infected larval or adult insects are blown by wind, hitchhike on vehicles, or travel on landscape plants such as sod. Conversely, infected but nonsymptomatic palms may be moved to new locations where uninfected insect vectors pick up the disease and spread it to other palms. Finally, plants that are not palms may act as incognito reservoirs for the disease, and they may be relocated and become accessible to insects that then move the phytoplasma to palms. That last scenario may seem remote, but the closely related Lethal Yellowing phytoplasma has been found in weeds in Jamaica, suggesting Lethal Bronzing may be lurking in plants no one suspects.[11]

In order to discern how the disease is transmitted by what vector, Dr. Brian Bahder was hired to replace retiring expert Nigel Harrison. Brian is an assistant professor of insect vector ecology who proved he could

identify a plant disease vector by working in California on a disease plagu-
ing wine and juice grapes. He notes that it was a lot easier working with
grapes in greenhouses than palms in the field.

Nigel had been a Florida plant pathologist for three decades, and when
TPPD showed up, he was on it. Many years before, probably in the 1960s,
thirty cabbage palms had been planted at the entrance to the Fort Lauderdale Research and Education Center. Then, in December 2014, after decades of stalwart service as examples of Florida natives that need little care,
five started declining. It was TPPD. These were the first cases anywhere
near the Center. This was the plant pathology equivalent of having the first
Ebola case turn up unexpectedly in your hospital. A fantastic opportunity
but also no doubt a little embarrassing. Plant pathologists had to watch
and document the collapse of their research center's entry landscaping. Dr.
Harrison and Ericka Helmick started testing the palms—painstakingly
recording their demise month by month. Numbers jumped from five to
seven, to nine, to ten, to fourteen. By May 2017, sixteen of the thirty cabbage palms had died. In June 2020, the number was twenty-one. The study
site could not have been more convenient, but it was a daily thumb in
the eye—an inescapable reminder that there was an unsolved puzzle that
threatened both Florida's native and exotic palms. The question of how
Haplaxius got there was especially vexing. Most likely it arrived on plants,
but no new palms had been brought in recently. Dr. Bahder suspects sod.
The fate of the remaining palms is unknown. Are they infected but not yet
symptomatic? Have they been exposed to the phytoplasma and are resistant as a consequence of good genes or general vigor? Or has the vector insect simply not made contact with the living tissues of the remnant palms?
The palms are dying to help Dr. Bahder find out.

Dr. Bahder came up with an ingenious experiment designed to clarify
how much sampling was necessary to detect the phytoplasma. He found
a dying cabbage palm at the research station, cut it down, and conceptually divided the trunk into thirty-three 6-inch-long segments. (After cutting the palm down he concluded that physically cutting the trunk into
6-inch segments was not worth the effort.) Then, for each section, he took
eight evenly spaced samples around the trunk, each rotated 45 degrees
from the previous sample. Finally, through a fairly laborious process, he
measured the amount of phytoplasma DNA in each of the 528 samples.
The testing revealed there was detectable Lethal Bronzing DNA in every

Cabbage palm entrance landscaping at the IFAS Fort Lauderdale Research and Education Center research station has been ravaged by Lethal Bronzing. Photo by B. W. Bahder—UF/IFAS.

sample, suggesting a single sample from a tree in decline should suffice for a diagnosis. But he also found intriguing, anomalous spikes in some of the samples, areas where there was far more disease present. Assuming the disease entered through the leaves near the top of the palm and then spreads throughout the plant, Dr. Bahder concluded these unusually high levels were associated with the virulent disease initially passing down from the canopy via the plant's vascular system phloem toward the roots, while the more moderate levels reflected the dispersed disease traveling back up toward the canopy, where it would eventually kill the bud.

By 2017 Lethal Bronzing had been in Florida for more than a decade, primarily as a concern for nursery growers, ecologists, preserve managers, and plant pathologists. The media found the disease newsworthy, and as early as 2006 an Associated Press/NBC story cited "a microscopic killer that has scientists stumped." It implied that budget constraints were likely to hamper looking for a solution.[12] Then, in the summer of 2017, the Tampa Bay Action News "I Team" broke a dying-palms story with an added government scandal twist. The investigation revealed that then governor Rick Scott had promoted "A Bold New Vision" for improved tropical (palm) landscaping along "gateway" highways. That led to planting expensive

Phoenix palms that started dying from Lethal Bronzing. Cue ominous TV news anchor: "You paid for them . . . pricey palm trees that line miles of Florida's interstates and highways as part of a Florida Department of Transportation landscaping program called 'A Bold New Vision.' But as the I-Team uncovered, many of the trees are now dead." The draw-your-own-conclusions implication was that the governor had promoted an ill-fated palm landscaping plan that benefitted a nursery that then coincidently poured fifty thousand dollars into the governor's reelection campaign.[13]

No one knows how problematic Lethal Bronzing may become. The insect vectors may be limited to grassy areas where the larvae feed—lawns and pastures, so palms in wild floodplain hammocks may be spared. Some palms may be immune and confer resistance to their offspring. Or the phytoplasma may mutate and become more virulent. As the magnitude of the threat becomes clearer, expect to hear more from Dr. Bahder and his team.

8

———— • ————

Groves

The Strange Relationship with Citrus

When we survey a landscape it is not unreasonable to mentally sort what we are seeing into threats and opportunities. And for those who manage landscapes in Florida, cabbage palms occupy a curious Jekyll-and-Hyde sort of niche. Urban and suburban landscapers are paid to import, plant, tend, and trim cabbage palms while later that day the same crew may be overpruning, killing, or removing dozens of cabbage palm seedlings from the same landscape.

Natural area preserve managers know well the wildlife values of our native cabbage palms, yet at least three preserve managers are waging war on our state tree. And pasture and grove managers frequently lose battles with invading cabbage palms, while others have been happy to have them.

Contemplating a declining grove, with fat, enthusiastic palm trunks pushing their striving green canopies up through ailing, branch-dead citrus is a poignant sight, one that reflects years of neglect and forebodes the end of the grove. Cabbage palms are not single-handedly destroying citrus groves, but many groves are vulnerable because a lapse in management reduces the likelihood that anyone will invest enough resources to remove the invading palms and restore the grove.

When the Spanish arrived in what they perceived as the New World, they brought with them every conceivable plant and animal that sustained them in the Old. Wine grapes, olives, apricots, date palms, and more. While some of their edibles survived, only three species really took off in the humid subtropics of Florida: cattle, hogs, and citrus. So while some of the plants that thrived in a dry Mediterranean climate foundered here, Asian citrus did well. Because animals, Native Americans, and settlers all spread the seeds, citrus became common, especially in what were known as low hammocks, a type of high water-table land dominated by trees such as live oak and cabbage palms.

Settlers curious about the agricultural potential of Florida created a market for books such as *Florida Fruits,* which was published in 1886. Among other topics, *Florida Fruits* featured a healthy debate regarding the relative merits of planting citrus in former hammock versus former flatwoods. Because citrus had naturally colonized low hammocks, settlers had little trouble concluding that hammock soils were the most appropriate for raising citrus. Unfortunately, most of the citrus trees found in hammocks were sour oranges, so some of the earliest production groves were so-called "wild groves," where landowners grafted sweet orange scions to the established sour orange rootstock. The resulting trees boasted roots well-equipped to deal with the low hammock soils and relatively high water table, yet produced sweet fruit for the market. The author of *Florida Fruits,* Helen Harcourt, was emphatic that the sheltering oaks in a wild grove should not be removed. Referring to these oaks as "Orange Guards," Harcourt mentioned a freeze "two years ago" and recounted the success of a grove on Orange Lake that had retained its oaks, while a neighbor who had removed his oaks "now declares he would gladly give twenty thousand dollars for a few of the stately forest trees that once sheltered his domesticated wild grove."[1]

Visitors to Highlands Hammock State Park can see wild (and wincingly sour) oranges growing among the oaks, but standing in the hammocks, it is obvious the oaks are undoubtedly competing for light and probably nutrients with the citrus. Someone at the University of Florida is rumored to have studied the role of canopy trees in protecting citrus from cold and concluded there was no real advantage in leaving the canopy trees. But people starting groves more than a century ago didn't have access to such analysis, and freezes are classic low-probability/high-impact events. Consequently, freezes were the one thing that could quickly destroy a grove, so some cautious people left oaks and others left palms, both to protect citrus from freezes. Camphor trees were sometimes left for good luck.[2] The serendipitous result was the creation of "some of the exceptionally good

This postcard probably dates from the 1920s and depicts remnant palms towering over an orange grove. Collection of the author.

coastal groves of the state. The old, so-called hammock groves were established with the undergrowth cleared out but many of the larger hardwoods and palmettos left in place, and were notable producers of fine crops of fruit. A few of these groves can still be seen along the upper east coast."[3]

Because a single live oak might conceivably shade a quarter acre, it was inevitable that the cost of land would push orange growers to optimize the number of trees they could plant and that the luxury of the sheltering oak limbs of the orange guards would yield to some other planting arrangement that didn't involve light-robbing competitors.

One challenge of a grove is creating conditions that favor the desired fruit trees while simultaneously creating conditions that discourage other, competing species. There are many modest annual plants and persistent grasses that plague groves, but while they compete for nutrients and water, they are not about to shade out the citrus. This year's crop of knee-high Spanish needles, if unchecked, will result in more Spanish needles, but vines, trees, and shrubs can be far more problematic if allowed to get a toehold because they can rob light from the citrus. The most problematic plants vary from location to location, based on nearby seed sources. Today exotic Brazilian peppers relentlessly try to invade many groves and can overtake the trees in just a few years. But historically and in groves adjacent to hammocks, seedling oaks and cabbage palms have been a reliable and persistent threat.

But suppressing plants around the periphery of the tree overlooks one problem, and that is competing plants that are able to get established underneath the sheltering limbs of the tree. Since many hammock species are capable of germinating and getting established in the low-light conditions of the hammock, they are quite at home under a citrus tree, and the shade found there is not an insurmountable obstacle. Cultivation under the tree is difficult because the lower branches impede access, and it's riskier because citrus trees typically have feeder roots close to the surface. The most dangerous place to cultivate, and therefore the safest place for a competitor to get started, is immediately next to the trunk. Wounding the trunk close to the soil line is an invitation to disease, so grove managers using mechanical cultivation are not eager to place steel blades in close proximity to intact trunks. So acorns and palm seeds fortunate enough to land near the trunk start with a distinct advantage. This fact becomes evident when grove management declines. With less rigorous cultivation, plants germinating next to the trunks get missed, while seedlings in the middles get cleared. I saw this firsthand when grove manager Peter McClure showed me a historic grove in western St. Lucie County. We drove through acres of geometrically perfect groves to enter a patch of hammock and an old grove that has been ravaged by cold, hurricane winds, and citrus greening, a bacterium that has been kicking the stuffing out of Florida citrus since 2005.[4] Many of the trees were missing, with just hints of the original rows, and those remaining were not in good shape. Despite occasional cultivation, the remaining trees frequently had cabbage palm sidekicks pushing up alongside them. If management is not restored to declining groves, the grove becomes unprofitable and is abandoned. This grove was too far gone to restore, but it is now being managed more like a game preserve, with feed strips and other fruit trees being watered. It was easy to visualize how, if the seedlings next to the trunks were to succeed, the result might be a hammock with native trees, either oaks or cabbage palms, in rows as straight as a nursery. The effect could be a forest dominated by native species aligned in rows, as if God developed an obsessive-compulsive fetish for order.

Of all the thousands of species and varieties of plants sold by nurseries in Florida, there is only one that is seldom propagated from seed or cuttings by nursery growers. Cabbage palms grow so slowly and are so abundant on wild lands that virtually no one can afford to collect seed, germinate it, and grow it to any significant size for sale. As a result, there are no abandoned cabbage palm nurseries where growers fell on hard times and abandoned their plants. Consequently, it might seem impossible to imagine a set of circumstances that would result in orderly rows of cabbage palms. But they are there, former citrus groves hidden in new forests of the east coast.

These linear palm patterns, though rare, are easy to see on aerial images such as the ones found on Google Earth, although they can be far harder to appreciate on the ground. In the most extreme cases, the replacement trees replicate the spacing of the original citrus. When the secondary replacement trees are cabbage palms, the result creates the appearance of an abandoned cabbage palm nursery.[5] But upstart seedling palms overtopping citrus are not the only palm-citrus relationship. A second, rarer still, relationship persists only in a few vanishing Indian River groves, relicts bypassed by contemporary grove management strategies.

The Indian River is less like a river than any other body of water called a river, with the possible exception of Marjory Stoneman Douglas's River of Grass. While it has banks and water, the Indian River is actually a 150-mile-long estuarine lagoon that keeps the main part of the Florida peninsula away from Merritt Island and the barrier beaches north of the St. Lucie inlet. It has no particular direction of flow and varies in salinity from freshwater to seawater strength. The lagoon is both beleaguered and remarkably diverse, but the name Indian River is famous not so much for the water as the region that surrounds it.

Champagne can come only from the officially designated Champagne region of France. Everything else is simply sparkling wine. Indian River citrus can come only from the officially delimited Indian River region. The most famous Indian River groves are on Merritt Island, best known for the Kennedy Space Center and the Merritt Island National Wildlife Refuge. Sadly, these groves are almost gone, victims of security concerns. But adjacent parts of the Florida peninsula are also allowed to brand their fruit as Indian River citrus, so there has always been an incentive to develop groves there, in the hope of capitalizing on the Indian River cachet.

South of New Smyrna Beach and Oak Hill, and west over the Florida East Coast Railway tracks and Turnbull Creek, is Turnbull Hammock, a forest that some settlers cleared to establish citrus groves. The towering cabbage palms outside of Oak Hill did not grow up in these groves but rather preceded them. These palms, remnant orange guards, were the only trees spared when the land was cleared for the grove. Of course, one has to wonder why a modern grove owner would tolerate other trees growing in the midst of their otherwise orderly rows. They must know they compete for nutrients, sunlight, and water. And their presence must complicate cultivation and other aspects of grove management, such as littering the ground with seeds that may eventually compete with the fruit trees. But both grove owners I've spoken with who own groves with emergent palms told me their experience suggests there may be something to the orange guards theory, reporting that recent freezes took a higher toll on adjacent conventional groves.[6]

Another factor arguing against their removal is inertia. It takes a lot of work to remove a tall palm from a grove. How do you fell it without flattening some fruit trees? And what do you do with the logs that can't be readily burned? One grower was fortunate to find a developer who would remove selected trees for free, but short of that, there is no cheap way to excise a mature palm from a grove.

Some of the oldest and most beautiful citrus groves are dominated by towering cabbage palms. These are the few remaining groves on low hammock land that have retained their orange guards. South Volusia and northern Brevard Counties are the places to look. Perhaps the best example extant is the Mullet Hill Farm between Oak Hill and Scottsmoor. The grove, planted in 1905, still sports dozens of mature cabbage palms. This was the grove so densely overshadowed that the owner had nearly twelve dozen removed to facilitate his management of the grove.

But even after the removal of dozens of trees, this modest grove presents a unique two-tiered aspect that is nearly breathtaking. This is the type of grove seen around the one-hour mark in the 1983 film *Cross Creek,* starring Mary Steenburgen as Marjorie Kinnan Rawlings.

Nearby lies what appears to be the third, and undoubtedly the rarest, relationship between citrus and cabbage palms. Imagine a grove owner a century or more ago leaving a few orange guards as he set out his small fruit trees. Now picture that grove thriving for decades but ultimately

The Mullet Hill Farm grove in Scottsmoor, Florida, evokes an earlier time when tall palms were tolerated or welcomed in citrus groves. Photo by the author.

declining—oaks, palms, Brazilian peppers, and other trees and shrubs invading and displacing the citrus. And yet, towering above the green fray are the persistent sentinels, the hammock palms that were spared and watched as the citrus came and went. They remain, the ultimate survivors, deliberately left for the protection they were believed to confer, then tolerated, and finally ignored and abandoned.

9

———— ◦ ————

Cabbage Palms in the Landscape

In April 2015, I drove 120 miles to attend a one-hour presentation to find out what landscape architects think about cabbage palms. It was a workshop at the third annual Native Plant Show of FANN, the Florida Association of Native Nurseries, a trade group that specializes in native plants, a challenge in Florida, where fortunes are spent to make the state look like the Caribbean, Pacific Islands, or even the Mediterranean.

The workshop was presented by Jamie Wright and Blake Gunnels of DIX.HITE + PARTNERS, INC., a landscape architecture firm from Longwood, Florida. Even though there are many species in the *Sabal* genus, "Sabals," it turns out, is Florida landscaper-speak for cabbage palms. They reviewed some history and various other uses of cabbage palms and then got to the point—their use in landscaping. They believe the average growth rate is about a half foot a year and that a 15-foot-tall tree might be around thirty-five years old. They showed slides featuring cabbage palms in both naturalistic settings and formal arrangements. They showed them as single specimens and as massed plantings. Searching for a possibly more objective take on this tree, I was most interested in what they thought the palm's strengths were. They listed five. The first of the five, perhaps counterintuitively, was "slow grower." Ordinarily, rapid growth in a landscape tree is seen as a virtue, particularly in the retirement communities of Florida, where impatient senior residents interested in timely, if not instant, gratification have been known to say, "I don't buy green bananas." Luckily the ability to transplant mature cabbage palms of any desired height obviates the need for them to grow quickly. This combination of features makes the cabbage palm ideal for urban settings, where they are unlikely to outgrow their designated roles. As a result, more than two dozen vendors advertise cabbage palms in Betrock's Plant Finder, a wholesale nursery trade publication.[1]

Wright and Gunnels also emphasized that it is a great wildlife plant, although they did not go into detail (cue slide with owl, raccoon, squirrel, robin, and woodpecker all perched in cabbage palms). They noted its cold tolerance (hardy to 15± degrees). They described it as "tolerant," which I took to mean tolerant of a variety of soils, drought (once established), and flooding. Finally, they touted its ability to work well in narrow and tough spaces, which may be attributed to the tolerance noted above combined with the fact that there is a reliable maximum size for the tree's canopy and the roots won't lift paving or pavers—you can plant them quite close to a building or in a narrow median. Then, several PowerPoint slides later, they listed ten reasons to design with native *Sabals*: Native, Iconic, Habitat, Adaptable, Survivor, Hardy, Resilient, Tolerant, Versatile, and Value.

High praise, and yet the state tree is not universally held in high regard in Florida. There's a planned community in the Orlando area called Baldwin Park where you won't see a single cabbage palm—they're not part of the approved plant palette. Technically, they are not banned outright, but they are only allowed in the "private" part of the yard, shielded from any street view.[2] The absence of conspicuous palms along the residential streets lends a temperate "Mayberry" ambiance. Some people simply don't like palms in general, others fear they harbor rats, and others don't like those droopy dead fronds. Baldwin Park's designers simply may have been accommodating northerners not yet ready for life in the subtropics.

The FANN Exhibition Hall was packed with displays featuring potted native plants tended by representatives eager to extol the virtues of their plant selections. I was drawn to the displays featuring live cabbage palms that had been dragged inside. There were five vendors supplying what are known as regenerated cabbage palms at the conference, and it is very impressive to see a sizeable palm tree growing out of a plastic-wrap container about the diameter of a truck tire. These trees dwarfed the selections of shrubs and groundcovers, and their presence was made possible by a relatively new approach to cabbage palm horticulture.

With the exception of Gary Hollar and some native nurseries producing seedlings for patient (or naïve) homeowners, cabbage palms, unlike other palms, are "field grown," which means they seed themselves on ranches

Regenerated cabbage palms overshadow other native landscape plants at FANN Native Plant Show. Photo by the author.

and grow for years untended until the rancher allows crews to come in and "pull" palms. For many years, cabbage palms were dug by hand or heavy equipment on such ranches, shorn of most of their leaves (the misleading "hurricane cut"), and the underground portion of the tree wrapped in plastic or burlap and kept moist until they could be planted. All the severed roots die, and a new set of roots has to be grown when the tree is relocated to its final site. If the palms had been (1) well hydrated to start with, (2) planted soon after being dug, and (3) kept moist while they grew a new set of roots, then they usually did pretty well. But transplant losses were higher than either the nurserymen or clients liked, and it took quite a while for the trees to recover and grow an entire new canopy—the stark "candlesticks" failed to convey the desired shady palm ambiance. When one or more died, it created an aesthetic problem and required the vendor to return with replacements—an aggravation for both parties. So there was motivation to increase survival of the transplants as well as delivering a plant that looked more like a healthy palm tree and less like a leafless stalk.

The solution is the "regenerated," or root-enhanced, Sabal palm. "Regenerated" primarily refers to the roots (although the canopy bounces back as well) and converts what could have been considered a defect (all the severed roots die when dug up and have to be replaced by a new set) into an asset (the opportunity to control the shape of the new roots in a manner

Recently planted cabbage palms that had been pruned for transplanting at Babcock Ranch, Florida. Compare with the regenerated palms in the previous image. Photo by the author.

that minimizes transplant shock). It transforms the traditional transplant-ing operation (dig-move-store-move-plant) into a new process: dig-move-reestablish-move-plant. It takes more time and care, which makes it more expensive, but transplant losses are minimized, and the customer gets a fuller, if not complete, canopy. The advantage of the regenerated palm is that it can be removed from the nursery and installed with little or no root mortality because the new roots (having been confined by plastic) are not cut when the plant is moved the second time.

The vendor with the biggest palm display was Steve Griffin, of Griffin Trees, and Steve is the man who perfected regeneration, or root enhance-ment. Talking with him was both humbling and refreshing. Refreshing be-cause he is so knowledgeable and humbling because Steve started dealing in cabbage palms when he was twenty-one, and he was fifty-four when I met him—so far, he estimated he had interacted with more than a million cabbage palms. He sells his trees from Virginia Beach to Texas. His wife and partner, Sheri, agreed that Louisiana has been a good market for them.

Steve and I disagreed about how far the roots extend. (I was taught at an IFAS palm workshop that the roots go out at least 40 and probably 50 feet—he wasn't buying it.) But in general our conversation was terribly affirming. We agreed that transplants do best when well-hydrated, that no one is likely to see a cabbage palm root bigger than the diameter of a pen, that an expanding trunk may shove pavers or bricks aside, but the roots won't lift them, and that (so long as you don't go through the pseudobark) cabbage palms could be sanded as smooth as anyone cared to.

Steve had wanted to try what became regeneration a couple of decades ago, but he was too busy selling trees to experiment. He told me he moved 120,000 cabbage palms into or out of his nursery in 2006. Then, when the economy slowed and the demand for landscape trees dropped off, he fi-nally had the capacity to experiment and gradually developed a technique for pulling the palms and optimizing their success. His teams don't dig palms so much as cut them out of the ground and then lift, or pull, them out. His heavy machinery uses a fixed blade to make first one and then another separate parallel cut on either side of the palm, and it then rotates 90 degrees to make another pair of perpendicular cuts as one might loosen a square section of cornbread with a knife. Then slings are attached to the trunk, and it is pulled out of the ground. His crews cut most of the green leaves off, leave the bootjacks on, and a laborer with a sharp spade trims the square root mass into a circle. What's left of the roots are wrapped in plastic, and the trees are driven to his nursery, where they are immediately misted. Then they spend six to eight weeks leaning on a tree rack, and when they seem to be recovering, the root area is wrapped in more plastic, the bottom plastic is cut off, and they are planted in the ground with persistent low-volume irrigation. I asked why he doesn't use pots. They're more expensive, and they can actually be harder to get off the root mass. His paradoxical goal is to produce a pot-bound plant, a plastic-encased ball of new roots seething in slow motion waiting to expand into real soil. His promotional literature summarizes the advantages as follows: "When our sabal palm is sold as a full headed root enhanced sabal, the palm is re-dug without disturbing the ball of the tree and the roots are visible through the stretch wrap. This rootball is now one big solid mass. This process is similar to a root bound potted plant. With plenty of water, the transplanting process is virtually fool proof. The customer now has immediate gratification and the landscape has eliminated the replacement problem."[3]

I also spoke with Will Ortelt from Boggy Creek Tree Farm, another outfit that produces regenerated sabals. He told me that, on average, ranchers get about ten dollars per tree (stumpage fee) for the palms they sell. He told me about trees with "iron boots"—ones with very persistent bootjacks that will run 20 or 30 feet up the tree. What does a sabal palm cost (the tree in the nursery, not delivered or planted)? In 2019, one could buy wholesale baby cabbage palms in one gallon pots for less than three dollars.[4] A regenerated fifteen-footer from Griffin was $240.[5] In general, the regenerated palms go for roughly twice what the unregenerated palms go for. Boggy Creek was selling regular field-grown trees for $110 and regenerated palms for $225. The process of creating a regenerated palm takes about fifteen months and is far more labor-intensive, but it speeds up the process of regrowing a complete, spherical canopy.

I argue for leaving all the fronds—living, dying, and dead—on cabbage palms, allowing the plant to adopt a natural, spherical canopy. My personal aesthetic favors a circular silhouette, and I observe that removing green fronds frequently replicates the aspect of a severely stressed or dying palm. According to the authors of *Ornamental Palm Horticulture,* removing dying leaves complicates diagnosis of nutritional problems because visual diagnosis involves examining all the leaves, especially the older ones. They point out that removing living parts of a plant is stressful and makes the palm more vulnerable to both insects and disease, as well as storm dam-age.[6] I also object to removing bloom stalks since the flowers are so impor-tant for pollinators and the seeds are important wildlife food. And while not everyone agrees it is a feature, the dead fronds can shelter northern yel-low bats, a common, solitary species that does not use bat houses, relying instead on "Sabal palm skirts and oaks with Spanish moss."[7]

Finally, since a five-year-old can safely drag a dead palmetto frond off the lawn, the costs implicit in the specialized equipment and professional skills needed to prune tall palms would seem to constitute an unnecessary expense fraught with risk and liability—particularly if a ladder or bucket truck is involved. I've seen a promotional video featuring a man holding onto a ladder with one hand while wielding a chainsaw in the other—the possibilities for mayhem are myriad. These are not theoretical concerns—in 2015, a man died trying to remove a palm in Ormond Beach.[8] In 2016 alone, one man died in Tampa while pruning cabbage palms,[9] and a tree service owner was killed while pruning a cabbage palm on Tybee Island.[10] In 2018 a landscaper in Jacksonville Beach fell off his ladder and died af-ter wasps attacked him while pruning a palm tree. He was fifty-seven.[11] In June 2018, a young man working a summer job was pruning a cabbage palm from a boom lift in Englewood, Florida. A damp frond fell onto power lines, shorting out power to his bucket, so he couldn't descend. He was electrocuted as he tried to escape. Forty percent of his body was burned, his legs were amputated, and he's endured "26 surgeries and 121 procedures." In 2019 he was learning to walk again.[12] Call me a cross between Don Quix-ote and Ralph Nader, but I don't think a seventeen-year old should lose his legs and some immeasurable part of his future because someone can't wait for a frond to fall.

According to the Bureau of Labor Statistics' annual National Census of Fatal Occupational Injuries report, those with the fifteenth-most-danger-ous job are "Grounds Maintenance Workers," and those with the ninth-most-dangerous occupation are "First-Line Supervisors of Landscaping, Lawn Service, and Groundskeeping Workers."[13]

Despite these compelling reasons to avoid pruning cabbage palms, over-pruning of cabbage palm fronds is a rampant behavior that appears to have multiple causes. Homeowner aesthetics, landscaper expedience, and a gen-eral suspicion that the palm frond bases "often harbor insects and rodents and may become a fire hazard" all seem to contribute.[14]

One crucial marketing point regarding the regenerated palms is the ability to install something with a canopy that approaches, or at least suggests, a normal palm canopy. The traditional approach to relocating cabbage palms involves cutting virtually all the fronds off, leaving something that has been called a candlestick and looks like giant, tan asparagus. This scalping is necessary because the rootless tree can't support many leaves. If assiduously watered, these stark columns eventually regrow roots and canopies, although I suspect the canopy of a relocated palm will never be as large as a cabbage palm that germinated in the same location and never been moved.

Residents moving into new subdivisions where the sod still has seams and the freshly planted, candlestick palms are nearly leafless may naively assume palms are supposed to be severely pruned. That may be one origin of the overpruned aesthetic. Landscapers are responsible for another.

Gone, it seems, are the days when some member of the household maintained the home landscape. In contemporary Florida subdivisions many homeowners hire landscaping services to "mow and blow"—deal with all the outdoor landscape exigencies some resident hailing from Michigan or New Jersey may not be equipped for, or interested in, handling. This typically involves mowing the sodded lawn areas (and dosing with fertilizers and biocides), trimming hedges, blowing fallen leaves into areas where they are less conspicuous, and pruning the palms.

Some palms are self-pruning. Royal palms, for instance, simply drop their massive leaves when they are done with them. But many species retain their dead, marcescent leaves. For some species, this accumulation of dead leaves is considered an attractive, positive feature, and the resultant encircling dead fronds are referred to as a skirt or petticoat. The California fan palm, *Washingtonia filifera,* and the Cuban petticoat palm, *Copernica macroglossa,* are both renowned for the distinctive volumes of dead leaves that typically exceed the number of living leaves. But palms don't need to be pruned—millions of wild cabbage palms do just fine with no pruning, yet some owners may not know that and assume their landscaper is just doing for the poor defective palm what it can't do for itself, like trimming a dog's toenails.

Despite the fact that palms don't need to be pruned, many people simply don't like the aesthetic of dead leaves hanging about. Ortho's home-owner-targeted *All about Palms* paperback advises that older, brown and dying leaves detract from a sabal palm's appearance but notes that "the best course is to remove palm leaves after they have turned completely brown."[15] Arborists and horticulturalists agree there is no harm in removing dead fronds, and dead frond pruning can be accomplished in several ways. Some shorter palms can be pruned from the ground with long poles, whereas others require ladders or even "cherry-picker" bucket trucks. Unfortunately, most landscaping operations don't stop at the dead fronds—they keep going, working their way up the palm, removing green fronds and bloom stalks as they go. Once in position, pruners tend to maximize their efforts, frequently resulting in less than a handful of green leaves remaining: "Experienced tree pruners have observed that the time until dead leaves reappear at the bottom of the canopy can be extended if they remove a number of living leaves from the bottom of a palm canopy while they are pruning off dead leaves. This is one of the justifications used for overpruning palms."[16] So some homeowners may simply defer to their landscaper's preferred approach, even though it is drastically reducing the plant's ability to photosynthesize.

A cabbage palm can have as many as forty living leaves. Leaving only five or ten means removing the majority of the photosynthetic capacity of the plant—something no one would contemplate for an oak or maple. One author states that cabbage palms produce five or six new fronds a year,[17] whereas the University of Florida estimates that a mature cabbage palm typically adds fourteen leaves a year, which agrees closely with a Seminole belief that they grow one new leaf each moon, for a total of thirteen a year. Whether the number is five or thirteen, it is easy to essentially erase a year's growth in a single pruning session.

Yes, those overpruned cabbage palms will probably survive. And that's where a cabbage palm's resilient strength becomes a liability. If these palms couldn't recover from massive leaf damage from wildfires, they wouldn't survive in lightning-prone areas. So a cabbage palm's remarkable ability to survive the loss of most fronds makes it vulnerable to a wide range of aesthetic and expedient approaches that should be considered abuse.

There is general agreement that landscapers should avoid the so-called hurricane cut, a tragic misnomer that suggests cabbage palms should be pruned in advance of an approaching hurricane. Ortho gently comments on the hurricane cut: "The practice has gone on for so long that many homeowners don't realize it harms the plant."[18] According to *Stormscaping,*

Although it is hard to believe, these resilient cabbage palms will survive wildfire and grow a new canopy. Note the standing dead pines killed in a previous fire. Photo by Maynard L. Hiss.

a book written in the wake of the terrible hurricane year of 2004, cabbage palms are the native trees most resistant to hurricane-force winds.[19] Ironically, there is evidence that leaving fronds on the palm actually works to protect the bud and that "overpruned palms were more likely to fail during hurricanes than those with their canopies intact."[20] So cabbage palms should be the last plant to worry about as a storm approaches. And overpruned cabbage palms have been shown to be more susceptible to cold damage.[21]

Despite professional agreement that the "hurricane cut" is counterproductive, it persists. In 2018, for example, eighty-five cabbage palms at Castillo de San Marcos in St. Augustine were overpruned by National Park staff ("in order to prepare for hurricane season"), prompting calls of outrage to the local TV station.[22]

The hardest sell is convincing people to leave a yellow or half-dead leaf. Why would a palm need a half-dead leaf? The answer lies in the fact that most palms in the Southeast are potassium deficient: "Under conditions of deficiency, the palm is able to extract potassium from the oldest leaves and translocate it to the newly developing leaves, allowing growth to continue in the absence of sufficient potassium in the soil."[23] So it is the dying leaves that enable the palm to recycle a critical nutrient.

I have come to accept the fact that it is nearly unavoidable to discuss palm pruning practices without reference to a circular clock face. People's familiarity with the hurricane cut may stem from seeing recently trans-planted cabbage palms, which are sometimes reduced to just the single upright spear leaf in the twelve o'clock position. Experienced landscapers willing to provide tender loving care and near-continuous irrigation some-times leave numerous leaves when transplanting with success, but more commonly only several green leaves are left. Based on the clock face anal-ogy, the hurricane or transplanting cut could be called the "eleven o'clock to one o'clock" cut, and some skilled landscapers with a commitment to irrigation adopt the "ten o'clock to two o'clock" approach for transplants, which still involves removing most of the living green leaves.

So what do the experts advocate? The University of Florida IFAS Exten-sion states: "A properly fertilized and pruned palm . . . should have a round canopy with green leaves right down to the bottom. Consumers must be educated that palms are supposed to have round crowns, not feather-duster crowns."[24] But Clemson and the International Society of Arbori-culture took a different position. Clemson Home and Garden Information Center: "Only remove completely dead and loose leaves, badly damaged or diseased leaves and fruit, and flower stalks when pruning a palm. If the petiole (the base of the leaf stem or stalk) is green, the leaf is not dead & never remove leaves at an angle above the horizontal (9:00 & 3:00)"[25] The International Society of Arboriculture (ISA) cited the American National Standards Institute (ANSI) 2008. After recognizing that fronds creating a dangerous situation should be removed, ANSI went on: "Live, healthy fronds should not be removed" followed by "Live healthy fronds above the horizontal should not be removed." So both sources simultaneously forbade and allowed removal of green fronds below a horizontal (nine to three o'clock) line.

The result was confusion regarding best practices. Fortunately, in 2017, the ANSI A300 Standard Practices for palm pruning were revised and the "nine o'clock to three o'clock" language removed. While ISA/ANSI stan-dards still allow yellow fronds to be removed for aesthetic reasons, aside from pruning for clearance, the standards do not provide any reason to remove healthy green fronds. Hopefully, Clemson's internally inconsistent recommendations (last updated in 2007) will eventually abandon the "nine to three" escape clause, which contradicts their admonition to not remove green fronds.

I believe there are only two situations in which removing green fronds is advisable. The first, as the ANSI A300 standards note, is for clearance. If fronds on shorter palms interfere with the movement of people or vehicles, or if the fronds are touching a building or threaten energized powerlines, it makes sense to trim. Second, cabbage palms are also pruned when transplanting, in order to compensate for the loss of roots. Conversely, if someone wants to weaken or kill palms, or has a generally disrespectful approach toward plants, pruning green fronds is one strategy.

Cabbage palms seem to be able to germinate and grow almost anywhere, pushing asphalt aside, alongside curbs, in the crotches of oak trees, anywhere a small seed can find something approximating soil. Ranchers and citrus grove owners find them aggravating invaders and have their own homegrown (some of which are probably illegal) strategies for coping with them.

Cabbage palms are banned, disparaged, pushed over, overpruned, cut down, and poisoned, but they keep coming—as inexorable as the brooms of the sorcerer's apprentice.

10

———•———

Picayune

Natural Florida Threatened by Native Palms

You won't find Mike Duever on Facebook. He's an old-school field guy and has been known to actually say things like, "I guess I'm more into ecology than people." He wears thick canvas-type pants even in summer to push through catbriers, poison ivy, or whatever defiant vegetation stands in his way. He's survived a venomous snakebite and a lightning storm that forced him to lie prostrate in a marsh to avoid attracting Thor's attention. He learned to navigate using a compass and the old USGS quadrangle maps and straps on a canteen and a machete before setting off on any adventure that will take him out of sight of his vehicle.

After a day with him you realize you've been spending it with a herpetologist who studied fish, which landed him a job with an outfit preoccupied with birds, where he pursued his curiosity about what factors governed what trees grew where in south Florida, which prepared him for his current job, which is focused on hydrology. He's not just a field person, but an all-around field person—his business card simply says "Natural Ecosystems," and after nearly a half century of observation there aren't many people who have studied Florida's landscapes more carefully.

When I met him, he'd spent most of eleven years as part of a team rejiggering the topography of 55,000 acres east of Naples, Florida, in an effort to create the hydrologic conditions necessary to reestablish towering cathedral-like cypress trees that preside over wetlands flooded much of the year. They've made great progress despite encountering a number of challenges: tightened budgets, manatees unexpectedly dependent on canal flows, possible lawsuits from neighbors who might be flooded, and cabbage palms.

The idea of restoring habitat suggests one can move backward to get back to some state that existed previously, but you can only move forward to a new state, a state that may be similar to a previous condition; you can't go back. Mike is a long-term thinker, and, since those are rare, he may be thought of as a visionary. And cabbage palms are definitely messing with his vision.

He's been working with the Army Corps of Engineers and South Florida Water Management District in a place called the Picayune Strand, although some still think of it as Southern Golden Gate Estates. Most of the merchantable cypress had been logged out of the area in the 1940s and 1950s, and then, by 1970, it became the world's largest lot-sale subdivision (173 square miles),[1] with 48 miles of canals and 270 miles of roads.[2] The scope of the land sale scheme is reflected in the fact that Mike's territory stretched nearly one hundred blocks from roughly Forty-Second Ave. SE to 134 Ave. SE. This quintessential Florida swampland sales operation lasted until 1974, when the owners declared bankruptcy.[3] By then Mike was already trying to improve the area, but it took years to acquire seventeen thousand parcels and remove 140 dwellings and tons of trash.[4]

By collecting data from a series of wells, Mike was able to prove the canals were lowering the water table as far as three miles away. The canals made the landscape drier for longer periods, which may have helped lot sales but led to the oxidation of organic soils, more frequent and severe fires, and allowed plants that had been stymied by wet conditions to get established. Plants such as cabbage palms. The solution was to scrape up most of the roads and push that material plus large quantities of excavated spoil along the canal, back into the canals to plug them. That and some very large pump stations will enable waters from the north to spread back out over the landscape, reestablishing the previous condition known as sheet flow, while maintaining drainage of surrounding private property.

As he drove us around some of the remaining roads in the Picayune, Mike was obviously proud of the changes he has overseen. Although it was dry season at the time, he took us near the eastern side of the preserve through what I would call a seasonally flooded savanna. Here the scattered

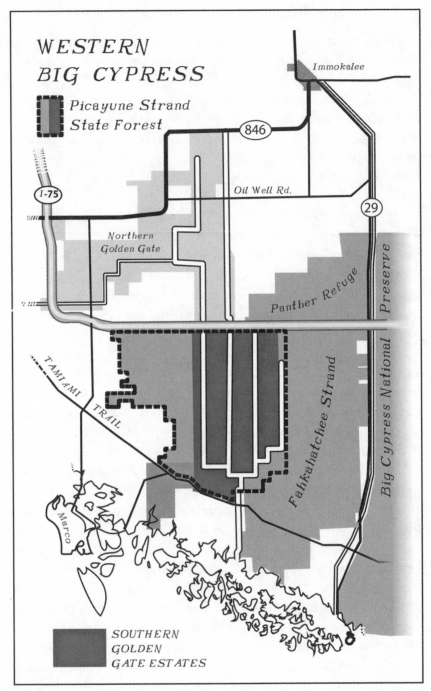

The western area of the Big Cypress Swamp had been extensively drained, creating conditions that favor invasion by cabbage palms. Map of former Southern Golden Gate Estates by the author, adapted from Chuirazzi and Duever, 2008 South Florida Environmental Report, Appendix & A-2: Picayune Strand Restoration Project Baseline.

trees were cabbage palms, thousands of them visible on the south side of 108th Ave. SE. Because the trees were separated and not clumped in one continuous closed canopy, it was easy to survey each tree as we passed.

"Stop the car." We had just driven past a cabbage palm that, even when viewed from a distance, was clearly having a bad hair day. The canopy was not a shaggy sphere but a busted-up collection of abused leaves. Most of the top half of the canopy appeared to be missing. A fire had burned through the savanna, leaving dark char on the trunks almost velvety in its light-absorbing blackness. We walked over to circle the tree I'd spotted and found deep parallel slashes about an inch apart in the char. I already knew that the charred trunks of palms were an excellent medium for recording claw marks. The passage of squirrels and raccoons is easy to distinguish by the width and spacing of the claw scrapes. This clearly was no raccoon, and since we had seen a bear crossing a road earlier, it was easy to conclude a bear had climbed this tree. But why this particular palm, as opposed to the dozens of others in the vicinity? Upon examination, it was shorter than some nearby alternatives, it had a clean, not booted, trunk, and it had seed-bearing inflorescences. Whether the bear climbed the tree to eat the fruits, or the heart, or just take a nap was unclear, but once we found this first tree we had no trouble spotting close to a dozen more that had been scaled by palm-loving bruins.

Although scattered cabbage palms were always found in the area, Mike believes most of the landscape previously had been too wet for the seedlings to gain a foothold. Mike's observation seemed to confirm what others had told me—that the seedlings are less resilient to flooding than are the older trees. Once established, these palms can tolerate a lot of flooding, so merely restoring earlier hydrologic conditions will not get rid of the existing palms.

The problem, as Mike sees it, is that the flood-proof palms also paradoxically happen to be highly flammable and also fireproof. The dry fronds burn like crazy, but the bud survives and subsequently resprouts. Since the area is typically dry part of the year in a region known for lightning, any invading palms that grow up in proximity to the remaining cypress constitute a conflagration waiting to happen, and so the remnant cypress that could conceivably hold on until the rehydration project is operational are being picked off by vicious, palm-stoked fires. It is tempting to hope the fires might set young palms back, but "In burning experiments, only

Florida black bear feeding on palm fruits after climbing 10 feet up into a cabbage palm. Photo by Tom Trotta.

seedlings less than three years old experienced any mortality."[5] So once the palms gain a fire-free toehold for a few years, they are functionally fire- and flood-proof. For Mike, then, our state tree is a "nuisance native" in Picayune.

The idea of a nuisance native is controversial and makes some Florida native plant advocates nervous. They promote native plants as less prob- lematic than those non-native plants that can spread uncontrollably into both managed landscapes and natural areas. The implicit promise is that native plants are better-behaved—that is, not invasive. But anyone with seed-producing cabbage palms needs to be prepared for seedlings gaining a foothold in planting beds, hedges, and next to any object that string trim- mers or mowers can't reach. As David Fox summarizes: "While cabbage palms add an inexpensive and low maintenance tropical accent to new de- velopments, they often 'pop up' and become established in existing urban landscapes within hedges and buffers."[6]

Mike Duever is not alone. Ecologist and fire expert Jean Huffman has studied and managed forest systems throughout Florida. She opined that if cabbage palms weren't native, they would be classified as invasive exotics. Ecologist Linda Conway Duever calls them "nativasives." Linda has been a stalwart Native Plant Society member since the first annual meeting, and she coauthored a detailed guide to successful cabbage palm transplantation. Conservation biologist Reed Noss, responding to a Facebook posting asserting native plants can't be invasive, wrote: "I disagree. 'Invasive' is properly applied to native plants that are weedy and respond to anthropogenic disturbance by invading native plant communities from off-site. This is accepted terminology. People incorrectly equate 'invasive' with 'nonnative' or 'exotic.' Not all nonnative species are invasive (in fact, relatively few are)."[7] It would be hard to find four people whose professional careers have been more committed to natural systems than Mike, Jean, Linda, and Reed. Yet, all four are quite comfortable calling cabbage palms invasive.

A 2013 Florida Native Plant Society blog posting led with the sentence, "Native plants are NOT invasive." The argument was that when native plants are "aggressive," they are merely the weedy species that colonize disturbed areas and will be replaced "after a couple of years"—early walk-ons in "Mother Nature's succession parade."[8]

Cabbage palms may show up relatively early, suggesting they are plants of early succession, but once established, they are persistent—these are not early succession plants that self-extinguish. Barring a palm disease epidemic, the palms are likely to be there for either a couple of centuries or until rising seas work their way up from Florida Bay.

Despite the claim that natives can't be invasive, the rest of the definition seems valid: a "plant that is expanding its range into natural areas and disrupting naturally occurring native plant communities."[9] To be fair, cabbage palms probably only move in when naturally occurring native plant communities are altered, either by natural successional phenomena or, more commonly, human practices such as drainage and fire suppression. And in 2018 the executive director of the Florida Native Plant Society wrote to assure me that "FNPS does not argue that native plants cannot be invasive! In reality, we must frequently consider and assess the invasive characteristics of native plant species in our restoration work."[10] The 2013 blog posting remained online, however, reflecting the pedestal some would reserve for native plants.

The danger in making native plants sacred is that it encourages a dualistic simplification that native equals good and exotic equals bad. That not only fails to prepare for the cabbage palm seedlings homeowners find springing up in odd places, but it also creates extra headaches for natural area managers who are trying to manage for pines, cypress, panthers, Florida scrub jays, or red-cockaded woodpeckers.

Mike wishes he could stop the palm reproduction, but I was unable to offer either solace or a solution. While it is true that new seedlings may be thwarted by restored water levels, those established cabbage palms may be in it for the duration. For a cabbage palm advocate it is sobering to see someone committed to the restoration of natural systems so vexed by our state tree. But that's just the tip of the nuisance native conundrum posed by cabbage palms.

11

———•———

Panther-Killing Palms

Disturbance Alters the Food Chain

Floridians know the panther has been having a tough time, but even knowledgeable residents might have been thrown off to see an October 2009 headline in the *Cape Coral Daily Breeze* that started: "Cabbage Palms Threaten Panthers."[1]

It is not immediately obvious how the state tree could threaten the state mammal—they just don't fall over that often. So how does a native tree, recognized as a fantastic wildlife plant, lead wildlife refuge managers to declare war on thousands of acres of cabbage palms?

For a while now the Florida panther has been vying for the unenviable status of being the most endangered mammal in North America. In fact, in the 1970s some experts believed there were fewer than twenty adults remaining, all in south Florida.[2] Other contenders for most endangered mammal include some bats, the wood buffalo, and the black-footed ferret, but anytime people are arguing about whether there are twenty or thirty animals left, you have a highly endangered species.

According to the Florida Panther Recovery Plan, this subspecies of the *Puma* faces numerous threats including habitat loss and fragmentation, inbreeding depression and reduced genetic health, vehicle strikes, intraspecific aggression, feline leukemia, pseudorabies, and mercury toxicosis.[3] That's a daunting burden for the popular and charismatic Florida State Mammal to contend with. Panthers are top predators, carnivores that rely on catching and eating mammals such as deer, feral hogs, rabbits, and mice. Without prey species, the other obstacles to recovery are irrelevant.

Historically those mammals did pretty well in the grassy wet prairies of the Big Cypress Swamp. These wet prairies were typically covered in water 50 to 150 days a year—conditions that prevented most trees from becoming established because the seedlings couldn't cope with growing in standing water. As a result, these areas were dominated by grasses, sedges, and wild-flowers, which are good foods for herbivores such as deer, wild hogs, rabbits, and mice. In addition to be being wet much of the year, these prairies typically had very thin organic soils over marl and limestone, which are known as calcareous (limey or alkaline) substrates.

As you can imagine, boosters with visions for the future of this part of Florida didn't think two to four months of inundation each year was a selling point, so drainage projects such as the Highway 29 canal were constructed to reduce flooding. With the wet taken out of the wet prairie, large expanses were made inviting to any plants that could cope with poor calcareous soil. Enter the cabbage palm.

The palms grew well, and that would have been fine if they provided food for deer, hogs, rabbits, and mice, but they don't. The leaves are tough compared to grasses, and the canopies grow up out of reach, creating shade that further reduces low-growing forage plants. The result is less prey for the panther.

The long-term solution is to restore the historic flooding regime, and managers are working on that, where they can. But there is a problem. Once the palms get past a seedling stage and develop a trunk, they can tolerate long periods of flooding. So part of the challenge involves beating back the palms, and there are a lot of them. In one place, managers counted 2,200 cabbage palms in one acre.

Florida Panther Refuge biologist Larry Richardson has two main tools at his disposal: Garlon 4 and the feller-buncher. Garlon 4 sounds like a Japanese movie sequel, but it's an herbicide that can kill cabbage palms. One ounce of his finely tuned Garlon 4 mix squirted in the bud will usually do the trick. The chemical alone costs two hundred dollars per acre, and there are more than 12,000 acres Larry would like to be able to treat. So far he has treated about 5,000 acres. The herbicide is used on low palms with buds within reach. One person can treat an acre with one thousand cabbage palms in three hours—they add a dye to the mix to distinguish which palms have already been treated (they really do all look the same).

Taller palms call for the feller-buncher, a product of the commercial forestry industry pressed into service for habitat management. Early on they tried clearing 200 acres with conventional chainsaws, and the feller-buncher is clearly superior—cabbage palms are surprisingly rough on chainsaws.

The feller-buncher is an arcane monster that has a horizontal toothed blade up front that is 3 feet wide and 1 inch thick. It spins at 600 revolutions per minute. The blade makes short work of the palm trunk while "tree clamps" ("giant malevolent claws" might be a better term) converge from either side to grab the tree and move it to one side. Because of its reach, a feller-buncher can clear a 23-foot-wide swath in one pass. One advantage of this machine is that it has low-ground-pressure tires, which exert only four pounds of pressure per square inch. That means that although palm trees have been ruthlessly sheared off and cast aside, the ground isn't all chewed up in its wake.

We head south on Highway 29 and get on I-75, Alligator Alley, heading west, then leave the interstate through a restricted wildlife refuge gate, and head north to investigate some study plots. Here Larry experiments with different approaches and uses game cameras to record the result. He has thousands of image captures of Florida panthers and notes that panther use of an area increases significantly in the year following a fire, presumably because the new growth and open aspect of the habitat makes life easier for herbivores and panthers alike.

Larry is in the position of having to eliminate palms, a role that has earned him some enmity from well-meaning native plant lovers who have a hard time understanding why the government is going after the state tree. Larry actually likes cabbage palms, and I sense he has been wounded over the years by people attacking the policy of removing palms. He is practiced at succinctly explaining why they can't be dug up and planted elsewhere.

Because Larry has been contending with this evolving problem for a quarter century, he has accumulated a provocative mix of observations and theories about what's happening, so as we roll along checking out different sites, I ply him with questions.

Why are the palms invading the pine flatwoods and cypress? The seedlings aren't drowning in the wet season the way they used to.

What's his goal or vision? His simple rule-of-thumb goal: Create the conditions so that alligators do not have to migrate. Right now their gator holes dry up, and they have to go look for deeper water. Then they end up overly concentrated in the canals and end up going after each other. The solution is a landscape that is wetter longer. He drives me past wetlands that are filling in with organic soil that he'd love to deepen.

What problems do the palms pose besides reducing the forage for the panther's prey species? The palms come in under the pines and cypress, and when fire moves through the area the heat from the burning palms can easily kill the pines and cypress. He's also concerned that the palms are drying out the landscape through evapotranspiration.

Can the pine and cypress areas of the preserve being invaded by palms ever be restored? He's optimistic. Get the hydrology back (which would drown the new seedlings) and then go after the larger palms that threaten the pines and cypress.

Has he ever seen cabbage palms killed by fire? He admits that's unlikely but then, upon reflection, qualifies his answer. If small palms are burned in the spring and that is followed by flooding, that combination can kill them because the palms are already stressed. Otherwise, no.

Leaving this remote site, to which I am unlikely to ever return, I ask to stop the truck to take a picture of dead, straw-colored, Garlon-ed palms. As we stop, we notice something unusual. I pick up on it and pile out of the truck. A tall palm had been growing next to, actually just a few inches from, a taller pine. The palm is dead and heavily charred, while the pine, also charred, is still alive. We try to reverse-engineer a plausible narrative since this state of affairs seems to directly contradict our combined experience and minutes-old consensus that mature palms are not killed by fire.

We speculate that when the fire came through, the remnant leaf bases (bootjacks) on the palm burned into the pine, which somehow, due to the unlikely extreme proximity of the trees, burned through the pine bark and may have set the pine sap on fire, which, in turn, caused more than a third of the palm trunk to be burned away. In any event, we were forced to reconsider our general rule about fireproof palms because the pine appears to have killed the palm, instead of the opposite textbook outcome we'd been discussing.

Some of the panthers' challenges have been successfully addressed. Mercury is less of a concern and the genetic depletion was addressed by introducing a modicum of out-of-state genetic material. But habitat fragmentation and vehicular collisions continue as significant problems. More panthers are known to have died in 2015 than were believed to exist in the

The palm on the right was apparently killed by the fire that the pine survived. Compare with photo on page 49. Photo by author.

1970s and 1980s—great news for optimists and depressing news for pessimists. And as if panthers didn't have enough problems, they are now feeding on domestic animals—primarily in Golden Gate Estates and frequently on goats, although they have taken calves, dogs, and ponies. These depredations, as they are called, create a public backlash that could threaten efforts to increase the range of panthers.

Like everywhere else, southwest Florida is ever-changing. Panthers will continue to hunt for deer and have trouble crossing roads. Cabbage palms will keep producing prodigious numbers of seeds that will germinate and become established unless efforts to restore a wetter landscape succeed. And Larry Richardson will retire, and someone else will take up the challenge of converting young palm forests into wet prairies that can better produce panther prey. And part of their job will be explaining why cabbage palms threaten panthers.

12

Birds and Bees

Wildlife in the Palms

At first glance, it may be hard to imagine what a cabbage palm has to offer animals. No sheltering branches, just a barren, pole-like trunk. A poofy topknot of maybe three dozen tough leaves. No traditional twigs to perch on and no succulent fruits, just modest fruits with hard, small seeds.

As it turns out, this simple combination is appreciated by many wildlife species and actually relied upon by some. I've found documentation of more than eighty species of animals using cabbage palms, not including oddball occurrences such as manatees feeding on the golden polypody ferns growing on cabbage palms that have fallen into spring runs, and giraffes noshing on lofty fronds in Florida wildlife parks.

Of course, familiar neighborhood animals use them all the time. Squirrels looking for nest material utilize the old bloom stalks for structure and covet the russet tufts of fibers found where the leaves join the trunk for the softer nest insides. Both squirrels and raccoons frequently clamber up into the crown to snooze curled up at the base of a substantial leaf stem.

*   *   *

It's early summer, and I'm walking through a landscape most would call devastated. The scrubby oaks and shrubs have been burned, and the cabbage palms have been crudely pushed over, apparently gnawed down by some enormous machine run amok in a protected Florida state park. It's probably just me, but the palms seem to have been singled out for destruction. From the damage I try to visualize its operation—imagining incessant steel teeth flailing through tough palm trunks. The now-flattened, open landscape creates a great opportunity to gain insight into the fibrous structure of their ravaged trunks, and I'm taking full advantage of it, taking lots of pictures and making notes about how it reminds me of shredded wheat—the big old-fashioned biscuits my father called monkey pillows, not their spoon-sized offspring.

I was last here with a college class in late spring when I made them wake up at (or stay up until) 6:30 A.M. on a Saturday and then had them walk in a perfectly straight line down an old railroad grade that is being converted into a recreational trail. Our guide that day was Sarasota County Commissioner Jon Thaxton, who in the course of 6 miles talked a lot about politics (including his longtime friend, former U.S. representative Katherine Harris), and also caught a black racer, a rat snake, and a ring-necked snake to show the students. Not your stereotypical county commissioner. Jon grew up in the area so he was able to place the landscape along the railroad in context, explaining that the family that owns the auto parts junkyard along the tracks is perfectly happy in their vocation and not likely to sell their 23-acre parcel, now estimated to be worth five or six million dollars. The students roll this around in their heads, and I sensed an emerging admiration for people who value familiar comfort over novel wealth.

We were nearing the end of our walk when, without any comment or explanation, Jon turned, crossed the railroad grade, walked several yards ahead of the group, and started making a repeated noise that sounded mostly like someone whispering, as loudly as humanly possible, the word "sheep."

"SHEEEP SHEEEP SHEEEP."

The students hung back, averting their gaze, unsure of what they were witnessing. Perhaps the commissioner had some embarrassing affliction? Or maybe he was a secret shepherd. Yet Oscar Scherer State Park seemed like an unlikely place to be calling sheep.

Jon persisted with his unlikely sheep call, although, awkwardly, nothing particular seemed to be happening, until a bird about the size of a blue jay launched out of an oak tree about 100 feet away, flew straight at Jon with alternating flaps and glides, and landed on a fence post roughly 2 feet from his head.

The foot-weary and sleep-deprived students were transfixed. Their initial sounds were subdued versions of the exclamations usually reserved for

Jon Thaxton
and wild Flor-
ida scrub jay.
Illustration by
the author.

firework displays. Finally, under his breath, one said, "He's the scrub jay whisperer."

While the bird cocked its head and studied Jon with a curious intensity, Jon delivered a short lecture about Florida scrub jays, interrupted only by the occasions the bird left the post to perch on the heads of some of the students who had worn straw hats. While the bird pecked at some of the hat ornamentation, Jon explained that the birds seem to favor people with blond hair or light-colored hats. Hatless students were profoundly envious while those with digital cameras captured multiple exposures of their peers with a wild bird enthusiastically whacking their heads.

I've heard his lecture several times, but it always intrigues me. Some of the key points are the pair fidelity of these birds—Jon explained how perfectly they exemplify Judeo-Christian values, never having been found to be adulterous; and the hard-to-believe statement that they can cache and relocate six thousand acorns a year, remembering not only the location but also the species of each acorn. As the students mentally tried to estimate the size of the bird's brain (ironically, it can't be any bigger than the acorns they bury), Jon told them western scrub jays are being used in memory and Alzheimer's research.[1]

Jon knelt down in the near-white sand to pantomime some of the steps the bird goes through in hiding an acorn, emphasizing how each separate behavior creates another possible memory. At one point the birds tap the location of the acorn with their beaks. I revealed to the students that instead of merely dropping my keys somewhere, I have taken to tapping the location several times, and it seems to help. On our way back to the van, we skirted the burned and leveled area I had come back to survey.

*   *   *

Someone had gone to a lot of trouble to savage every cabbage palm of any size and then burn the area. And, indirectly at least, that person was Jon Thaxton. Upon inquiry I learned that park managers were targeting several tree species for removal and the palms were the biggest trees standing in the way of their goal, which is to improve scrub jay habitat, which in Oscar Scherer State Park is a habitat variant known as scrubby flatwoods. Pine flatwoods are the most extensive terrestrial ecosystem in Florida, and in some areas they possess better-drained soils with more scrubby oak species in the understory and more open, sandier conditions at ground level.[2]

For the jays, scrubby flatwoods habitat is the zenith of habitat conditions, and anything that distorts the structure of the scrubby flatwoods endangers the jays. That structure includes relatively low, dense oak vegetation, with intervening open, bare-sand areas. When the vegetation gets too high or too dense, the jays decline. Historically, lightning-induced fires would rip through this habitat, consuming the aboveground portions of the oaks—resetting the system every few years to provide ideal scrub jay habitat. But while the number of lightning strikes may have remained constant, the acreage burned each year has been vastly reduced due to both firefighting and the fragmentation of habitat, frequently by roads, which function as firebreaks. The result was that in the absence of fire, Oscar Scherer State Park, renowned for its jays, experienced a significant decline in the number of jays.

Thaxton, who had been studying the jays for years, noticed that they could be found in unexpected areas where dense vegetation was juxtaposed with open and cleared cuts, such as the area under a powerline. That observation made him wonder if artificial clearing could make the possibility of reintroducing fire more feasible because a low layer of fuel burns differently and is more manageable than standing fuel such as scrub oaks.

He envisioned a two-step process in which heavy machinery would knock the oaks and other plants to the ground where they could be burned to create the open structure jays seemed to seek. The first experiments at Oscar Scherer were controversial, radically unproven attacks on native plants in a state park. While park users questioned what the hell was happening to their park, state park administrators openly wondered if it would, or even could, work. After holding their breath through the conflagration, park managers were relieved to see jays entering the area to look for singed insects while it was still smoldering. It turns out that it takes three or four years following a burn for the oaks to be dense enough to support nesting, but Thaxton's insight and radical treatment made it clear that crash-and-burn habitat restoration might work.

The area I was inspecting had been colonized by numerous cabbage palms, which had slowly degraded the habitat, at least from the scrub jay perspective. So the palms had to go to make the area suitable for scrub jays. The irony is that, here at least, scrub jays may actually depend on cabbage palms for their survival.

At first it isn't obvious why an omnivorous bird that runs an acorn-based food storage operation might need palm trees. Palms do not produce acorns. And palm seeds aren't known to be a significant part of the scrub jay diet. They don't nest in palm trees, and too many palms actually render the habitat unsuitable.

The answer lies in the nests, tucked as they are in the dense scrub oak. Scrub jay nests are about what you would expect for a bird the size of a blue jay. They would fit in a cereal bowl and are two-layered constructions. The outer structure consists of dead oak twigs gathered from branches on live trees. In other words, the jays don't hop around on the ground picking up fallen twigs but harvest their own from dead branchlets. Since stiff oak twigs aren't too different from toothpicks, it is easy to understand why a different, more chick-friendly, inner layer is needed.

The inner layer, the nest lining, is composed exclusively of the fine, hairlike fibers that protrude from between the fringed margins of mature cabbage palm leaf segments. The blond fibers are some of the last things people notice about cabbage palms. Many descriptive texts fail to mention them, yet they are diagnostic and simplify the task of distinguishing small cabbage palms from robust saw palmettos. When I ask people what these nearly anonymous fibers are called, I typically get personal, anecdotal names like "monkey hair." Botanist Scott Zona calls them "filamentous fiber extensions."[3] Until noticed and brought to consciousness, the fibers are nearly inconspicuous, even though they can be more than 2 feet long.

These stray wisps are remnants of the palm leaf segments, which split as they move away from the center of the frond. The fringy ends of the leaves that lend chickees and tiki huts their character are leaf segments, and each leaf segment has a strong supporting rib. The fringe effect arises as the leaf blade divides into leaf segments and the remnant rib projecting between leaf segments becomes the "monkey hair" fiber, dangling in the space between two adjoining leaf segments. On a cabbage palm each leaf segment is divided once again, and frequently each minor rib then sports a shorter monkey hair. Once freed from the leaves, the fibers frequently hang in solitary disarray, but loose curls and even open Shirley Temple ringlets are also seen.

I've seen students, inspired by Native American lore, collect monkey hair to make cordage; the fibers are remarkably strong, and some swear by it as tinder. But other than rope-making and fire-starting, the only known use for these palm fibers that I have come across is nest building.

Elsewhere, in cabbage palm–poor locales, the scrub jays have other strategies, other fixes. Scrub jays on Florida's Central Ridge use comparable fibers from a closely related palm species: *Sabal etonia*. And the jays fortunate enough to still have viable habitat near beaches use fine root fibers, which are exposed as dunes erode. But here in the scrubby flatwoods of Oscar Scherer, monkey hair is the only known nest-lining material, and it is apparently so crucial to nest building that the nesting pair will leave their home territory (a big deal in the world of scrub jays) to collect the fibers elsewhere.

The problem with leaving the territory is that they usually have to cross another jay family's territory, subjecting themselves to a gauntlet of harassment from the neighboring jays. If they choose to fly higher, gaining altitude to avoid the neighbors, then they risk attack from either sharp-shinned or Cooper's hawks. Yet they risk these challenges repeatedly to seek out these must-have fibers. They achieve some efficiency by collecting three to five fibers at a time, and, having gathered them, they frequently perch with the fibers between their claws and pull them through their beaks. No one knows if this is to clean them, soften them, test their tensile strength, or what; but once they're done they head back to the nest site.

The first nest of the spring can take two weeks to build and involve a lot of fibers.

How many fibers do they need for one nest? I naively decided I was willing to conduct some basic palm utilization research, so I asked Jon if he could secure a used scrub jay nest for me. Since the jays don't reuse nests, he couldn't see how it would create a problem for the birds if he recycled one. After a season or two I obtained an abandoned nest from Nancy Edmundson, who managed the Manasota Scrub Preserve for Sarasota County.

The nest arrived in a cardboard box, having been pruned free from its oak, and was roughly 10 inches across and constructed of tough, wiry oak twigs. The juxtaposition of the angular twigs enclosing the soft curves of the palm fibers was striking. My father once found a bird's nest lined with white horse hair, and this scrub jay nest reminded me of that. It soon became apparent that the nest was really two separate structures, one embedded in the other. The several dozen twigs formed spiky scaffolding that contained the swirling, cushioning palm fibers. Although some of the fibers were snagged on stiff twigs, I concluded the inner nest was built as a separate second phase, and I was able to extract the inner nest, leaving only three dozen fibers snagged in the twigs. The inside of the nest perfectly held a tennis ball, and the thickness of this inner nest wall varied between and inch and an inch and a half.

My approach was to count ten long fibers, make a note, and then return to the nest. I quickly found that there were numerous short fibers, presumably fragments of longer fibers broken down by nestlings or merely age. I didn't have the patience to straighten each of the curved fibers so I don't know what the average length was, but I started counting several short (3- or 4-inch) fibers as one long one. I proceeded from the center of the nest outward. It wasn't long before I had a sense of how much tension I could apply before a strand was likely to break. The nest seemed remarkably clean, but I was working on a table where we eat, so I paused to put down a layer of newspaper. After 100 PFEs (my newly coined Palm Fiber Equivalents), the nest looked very similar in diameter to what I had started with, but a little thinner on the bottom. That allowed me to postulate that, for this nest at least, the walls preceded the thickening of the floor. After 130, I started wondering about repetitive motion injury but remembered the small birds had to do all this in reverse, and I pushed on. After 150 the nest

Spiky oak twigs contrast with the soft palm filament lining of a Florida scrub jay nest. Photo by the author.

was dramatically thinner on the bottom, but the walls still seemed mostly intact. I started feeling confident there were at least 500 fibers involved.

I took a break for the night, and the next morning I witnessed an osprey bringing a slender 3-foot-long branch to a nest it was attempting to build in a dead pine snag in our front yard. Their nest scaffolding is impressive, and the lining is Spanish moss, which can be seen dangling from their talons as nest-making proceeds. One does not need to know the relative contributions of learning and instinct to be impressed by nest building, an important, but very occasional, activity that seems generally unrelated to the nonbreeding existence of birds. I resumed my fiber extractions, and did make it to 500, with fragments possibly equaling another nine, depending on what the minimum length a jay would be interested in actually is.

Based on what Jon Thaxton had told me, I calculated the nest I destroyed required between 100 and 150 trips, with many probably through hostile neighboring jay territory. Impressive.

Would the scrub jays of Oscar Scherer drift toward local extirpation without nearby cabbage palms and their filamentous fibers? No one knows. Jays are members of one of the smartest groups of birds and might adopt substitutes if forced to. They use other material elsewhere. But their willingness to leave their home territory in search of these fibers suggests they don't perceive suitable alternatives close by. And despite their intelligence, their inability to adapt as scrub disappears suggests there are hard limits to how much they can adjust. The challenge of managing habitat for listed species is epitomized by the fact that these rare and distinctive birds may actually depend on fibers so inconspicuous that poorly trained managers could easily overlook both their existence and their importance.

Florida scrub jays are just one species of bird that utilize cabbage palms. Other birds no doubt use palm fiber for lining their nests as well. The most prosaic use of a palm is simply as a perch, and while I've only found photographs of thirty bird species using cabbage palms as perches, the actual number is no doubt far higher.

What is intriguing is not the perch use so much as the other, more specialized, uses. Audubon's caracaras nest in live cabbage palms,[4] and ospreys sometimes build nest platforms on dead cabbage palms.[5] Vultures roost on them, breaking down the leaves with their weight.[6] Nest-making hummingbirds are said to use other cabbage palm fibers that occur where the leaves join the trunk, but so far I've been unable to confirm this.[7]

Florida has eight species of woodpeckers, and all but three (red-cockaded, hairy, and the red-headed) are likely to be seen working on cabbage palms. The smallest species, the downy, works the leaves and bootjacks and can make quite a racket hammering away.[8]

Migratory sapsuckers have been known to riddle the trunks with round, oozy holes to create gummy feeding stations. I've seen these most often where the trunk is constricted or necks down for some reason. Flickers and red-bellied woodpeckers can also be found on cabbage palms, but the most dramatic are the pileated woodpeckers, which frequently nest in dead palms. Some standing dead palms have multiple woodpecker holes and look like immense standing flutes. And swarms of honeybees have been known to occupy old woodpecker nests.

Cabbage palms bloom during the summer, not in one well-defined burst, but in a rambling cascade over a number of months, sometimes into October. The flowers themselves are quite small—one could easily hide behind a lentil. The flowers are found on long panicles (bloom stalks), and I cut a panicle in my yard and brought it inside. At more than 8 feet long, it overwhelmed our kitchen table. My panicle had twenty-two branchlets, each shorter than the one before. The flowers aren't located on the branchlets but rather on slender stalks called rachillae that are found along the branchlets and that can be up to 6 inches long. There can be as many as twenty-two rachillae per branchlet and more than fifteen flowers per inch along the rachillae. The longest branchlet, the one closest to the trunk, was nearly 30 inches long and held ten rachillae. I counted the flowers on each for a total of 4,865 flowers. I didn't have the patience to count all the other flowers, so I counted the total number of rachillae on all branchlets (1,747) and assumed the length of the average rachillae was 3 inches. At a conservative fifteen flowers per inch, that yielded 78,615 flowers. That seemed hard to believe, but Kyle Brown calculated that there were more than 85,000 buds on one panicle he studied.[9] Kyle Brown states that six to eight panicles per palm is a common number, but the palm by my driveway had only two blooming panicles. On the other hand, I found a cabbage palm less than 10 feet tall with nine current panicles, and it had been producing bloom stalks since it was knee-high. Another had seventeen contemporaneous panicles.

So with six comparable panicles, a single palm might sport a half million flowers. Since a palm hammock could have as many as nearly four hundred mature palms per acre, we're talking about a theoretical flower potential somewhere around 180 million flowers per acre. Even with a minuscule amount of nectar per flower, that is a tremendous resource for nectar- and pollen-seeking insects, a list that includes seven species of beetles, six species of flies, three species of true bugs, a treehopper, and more. But the group of insects that is really attracted are the forty-five species of hymenoptera: the bees, ants, and wasps. Of these it is the bees that predominate. So many species of bees frequent cabbage palm flowers that wasps are intimidated.[10] Kyle Brown found two dozen species of bees frequenting cabbage palm flowers, and, of these, honeybees are especially common, perhaps because, unlike most bee species, honeybees are semi-domesticated and consequently abetted by humans. So not only are wild colonies sometimes present in hollow trees, including cabbage palms, but some beekeepers deliberately move their hives near palm hammocks to take advantage of the summer palm nectar flow. According to Brown, bee species harvest both pollen and nectar and the nectar constitutes "a major source of food" for various bee species.[11] Beekeepers have learned that honeybees eagerly seek out the cabbage palm nectar. Bees collect flower nectar, thicken it through regurgitation and place it in hexagonal wax comb, leave the cells uncapped, and use their wings to hasten evaporation. Once it is thick enough, they cap it for use later or until a beekeeper slices the caps off and extracts the exposed honey.

It is not too hard to find a product labeled palmetto honey, which many people regard as high quality. Yet "palmetto honey" is almost inevitably not cabbage palm honey but rather saw palmetto honey. Curiously, the cabbage palm nectar the bees collect is problematic. In the 1909 volume of *Gleanings in Bee Culture*, L. K. Smith wrote, based on his personal experience on the east coast of Florida: "Cabbage palmetto honey, sealed or unsealed, will foam as though fermentation was in progress; that taken from the combs unsealed will ferment enough to deprive it of all honey flavor, but the sealed only foams. Thin and acid and amber in color, it will flow bubbling from the cells behind the knife and it is not a rare thing to see gas bubbles under the cappings of the sealed cells. Whether the colonies are strong or weak, it is always the same, when the bees work on cabbage trees, as the common palm tree of Florida is called."[12]

Even once thickened and regurgitated by the bees, cabbage palm honey remains watery, so watery that yeasts naturally found in the environment can multiply and initiate fermentation. Matured honey doesn't have enough water to support fermentation and lasts for years, but thinner unmatured honey can ferment, and the regurgitated nectar can actually start fermenting in the hive. Perhaps that's because cabbage palm blooming coincides with the humid, rainy summer months that can hamper evaporation. Syrupy, water-laden, unripened honey can't be sold as honey until it is less than 18 percent water because it runs the risk of turning into a foamy

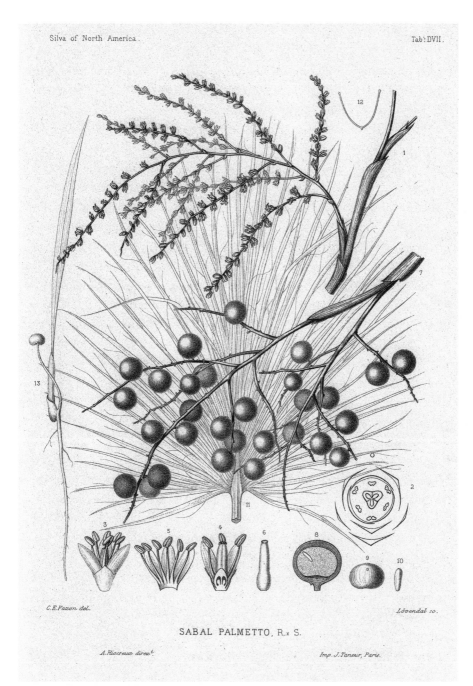

SABAL PALMETTO, R. x S.

Sabal palmetto flowers, panicle, and fruit. Engraving from *Silva of North America,*
vol. 10, by C. S. Sargent (Boston: Houghton Mifflin, 1896), plate DVII. Author's
collection.

natural form of mead. Anxious beekeepers may pull the thin cabbage palm honey before it can ferment and try a variety of ingenious heating, fanning, and other dehumidification strategies to reduce the water content. If it remains watery, they may be paid less by wholesalers who then blend the cabbage palmetto honey with other thicker honey.[13]

Even when it is dried-out enough, you're unlikely to buy pure cabbage palm honey because lots of other wildflowers such as Spanish needles (*Bidens pilosa*) are blooming at the same time, so unless the hives are placed where nothing else significant is blooming, it will inevitably be a mixture. Since it is problematic and seldom pure, you don't often see cabbage palm honey promoted like tupelo, orange blossom, or clover. However, some entrepreneurial beekeepers have tried to create a niche market for cabbage palm honey. Usher Gay, an educator with the Savannah Bee Company, posted a blog entry in 2015 describing *Sabal palmetto* honey as "the very best honey in the United States." He characterized it as having "a rich smokey flavor that is sweet and robust." Mr. Gay mentioned its antioxidants and vitamins, recommended it for iced sweet tea, and concluded with: "I look forward to the harvest of this honey more than any other during honey season. I hope that you will experience this honey for yourself. I would be willing to bet that it will not disappoint and it might even become your new favorite honey!"[14]

Naturally, I wanted some—how often do you get to taste a honey with a greater reputation than tupelo? It took a while, but I found a source of cabbage palm honey from the Turner family of Cedar Key, Florida. I bought a pint and a half for eighteen dollars, which was less than half the cost of Savannah Bee's. The owners confirmed that they kept their honey house hot and ran a dehumidifier, but did not cook or blend the honey. The honey was very light in color even though it was in a cylindrical jar and not one of those laterally compressed jars used to feature the glowing radiance of honey. The Turners told me it was a particularly sweet honey, which seemed improbably redundant, but it turns out honey contains a variety of different sugars and some are sweeter-tasting than others. The predominant sugar in cabbage palm honey is fructose,[15] which is 1.7 times sweeter than sucrose (table sugar). There is more fructose in cabbage palm honey than all the other major sugars (sucrose, glucose, and maltose) found in cabbage palm honey.[16] Setting aside the fact that I am completely biased, it did seem remarkably sweet, although I failed to detect any "rich, smokey flavor." It had none of the aftertaste I detect with black mangrove or tupelo, and was much better than sea grape or avocado.

In keeping with the cabbage palm's paradoxical tendencies, opinions vary regarding whether cabbage palm honey is suitable only for getting a colony of bees through the winter, or it is a desirable table honey. In 1954, Lillian E. Arnold published *Some Honey Plants of Florida,* a work that dealt only with native plants, and consequently only five plants were listed as yielding commercial quantities of honey: saw palmetto, partridge pea, gallberry, Ogeechee tupelo, and black mangrove. Cabbage palm honey was described in a section titled "Contributing to Colony Maintenance" and "is said to be thin-bodied, with a high moisture content and has the added disadvantage of tending to ferment after extraction unless heated immediately. However, this honey serves well for winter stores."[17] In 1946 the *Tampa Tribune* claimed that "cabbage palm honey is a thin-bodied, light amber honey of very mild flavor and odor. It is excellent for cookery and sweetening drinks where mild flavor is desired.[18] Those desiring to arrive at the truth will have to obtain their own source of this pale, thin honey and draw their own conclusions. Bring your wallet—these are not good times for honeybees, and cabbage palm honey can be considered rare, if not downright artisanal.

The broad arching leaves of a cabbage palm provide a green umbrella well-suited for protecting the little paper wasp, *Mischocyttarus mexicanus cubicola.* These vigilant insects are not particularly aggressive, but since their compact nests hidden under fronds are seldom seen by lumbering humans, most people don't notice the wasps' head-on warning stare before they are provoked enough to fly out and start stinging.

It's not just wasps, bees, and Florida scrub jays. Consider some of these more unlikely palm aficionados: Several species of turtles in rivers and lakes rely on fallen palm logs to sun themselves, lining up one after the other to bask. Baby alligators do the same. Northern yellow bats, *Lasiurus intermedius,* shun bat houses (and attics) to roost almost exclusively in dead cabbage palm leaves or clumps of Spanish moss.[19] Florida panthers, not having access to your living room couch, claw-mark fallen logs, especially cabbage palm logs. This behavior involves leaving scent markings from glands between their toes.[20] Florida black bears rip into young palms to eat the hearts and then sometimes return to eat the palmetto weevil larvae that grow in the decomposing heart remnants.[21] The Gulf salt marsh mink

2009-08-01 1:13:56 AM   M 1/3   71°F

FWC03000455   **Male panther claw marking a log**   RECONYX

A male Florida panther claw marking a fallen cabbage palm log. Photo by Roy McBride. Image originally appeared in "Photographic Evidence of Florida Panthers Claw Marking Logs," *Notes of the Southeastern Naturalist* 10, no. 2 (2011): 385.

(*Mustela vison halilimnetes*), one of the rarer and more elusive carnivores in Florida, hides out in hollow cabbage palm logs.[22] Rat snakes use prodigious free-soloing climbing skills to ascend even smooth-trunked cabbage palms in search of bird nests in cavities. Anoles lay their eggs in the accumulated detritus in the leaf axils and, sadly, Florida box turtles occasionally fall into rotted out palm stumps (natural pitfall traps) and succumb.

Between living and dead palms, standing and fallen palms, hollow and solid trunks, seedlings and canopy trees, flowers and fruits, it's hard to imagine a species of terrestrial southeastern wildlife that has not interacted with cabbage palms in habitats where they both occur.

13

———•———

Coastal Creep

Natural Distribution, East Coast

Key West has a reputation as a wild and rowdy place, but courtesy is reliably on display during daylight hours at the corner of Whitehead and South Streets, where tourists queue up to have their pictures taken at the odd Southernmost Point monument. There one can frequently find a line of people politely waiting as those who got there earlier, snap, hand off cameras or cell phones, and trade places. Despite the hype surrounding the monument at the street corner, parts of Fort Taylor to the west are actually farther south (but are not publicly accessible), and no one pays the slightest attention to the nearby southernmost cabbage palm. Which makes sense, because the southernmost cabbage palm with the trunk scuffed-up from bicycles being locked to it is only the southernmost cabbage palm in the United States since *Sabal palmetto* naturally occurs outside the United States. The southernmost naturally occurring cabbage palm is somewhere in Cuba, and the easternmost in the Bahamas.[1]

Geographic range limits are intriguing, and botanists enjoy discovering these extremes, particularly when they represent natural range extensions and are not the result of horticultural transplants, such as the planted palm in Key West.

Kyle Brown wrote his doctoral thesis in 1973 on the "Ecological Life History and Geographical Distribution of the Cabbage Palm, *Sabal palmetto*," so consequently he occupies a special, and spatial, place in the pantheon of cabbage palm experts. Brown's insights were informed by his studies of the spotty distribution of cabbage palms north of North Inlet, South Carolina.

Cabbage palms are common on the barrier islands of Florida's east coast and Georgia's sea islands. While this book focuses on the two states that saw fit to designate the cabbage palm (Florida) and palmetto (South Carolina) as their state trees, coastal Georgia merits attention. Starting in the south, the palms that get buried and subsequently reemerge from Cumberland Island's massive migrating sand dunes are some of the most dramatic survivors anywhere. The historic avenues of palms on Jekyll Island are some of the oldest North American landscape palms known. At the south end of St. Catherine's are some intriguing fingers of palms growing on relict dune ridges. Ossabaw Island has both gorgeous palm savannah and hundreds of severely stressed and headless, standing dead palms, victims of rising sea level. And there are thousands of happy, healthy cabbage palms on Wassaw, an island that seems remarkably stable for nowhere being more than three-quarters of a mile wide. I don't think any other 100-mile stretch of coast has more cabbage palm diversity than coastal Georgia.

Starting in South Carolina and moving up the coast, cabbage palms (now called palmettos) can be reliably found on Dafuskie, Hilton Head, Bay Point, St. Phillips, Pritchards, Fripp, Hunting, Otter Islands, Edings, Pockoy, Seabrook, Kiawah, Folly, Morris, Sullivans, Isle of Pines, Dewers, Capers, and Bull Islands. Then there are a series of low-elevation, frequently overwashed, islands with no palmettos. The next wild palmettos seem to show up just north of North Inlet about 38 miles north of Bull Island on the north and south ends of Debidue Island. Brown thought these were the northernmost palmettos in South Carolina, although there may have been relict palmettos on heavily developed Pawley's Island to the north.

According to Kyle Brown, there were at least two additional natural populations farther up the coast that were "completely removed for landscaping or otherwise destroyed." The strand along Myrtle Beach is heavily developed (and heavily planted with introduced palmettos), but aerial review of predominantly natural Myrtle Beach State Park reveals no wild palmettos. Brown argued the climate was not materially different along that stretch of coast and instead observed that cabbage palms didn't become "established on beaches but rather on the bay sides of islands and adjacent mainland."[2] Since the section of the South Carolina coast in question is primarily barrier beach (locally called strand), and not true barrier islands, Brown concluded that cabbage palms are "absent for the northern South Carolina coast because of the lack of habitats suitable for germination and seedling establishment."[3] Brown believed palmettos are primarily water-dispersed and consequently end up on the bay side, not the beach side.

The northernmost naturally occurring cabbage palms lie in North Carolina. In 1819, F. A. Michaux wrote that palmettos were "first seen about Cape Hatteras, in the 34th degree of latitude."[4] The actual Cape is above 35 degrees, so he may have been speaking imprecisely, which would place his first cabbage palms in the vicinity of Bald Head Island, farther south. Bald Head Island is not technically an island—one could walk there from Kure Beach, but it is functionally an island since it is normally reached only by private boat or ferry from Southport.

The vegetation on Bald Head Island would seem familiar 400 miles farther south: live and laurel oaks, beautyberry, wax myrtle, southern red cedar, poison ivy, cabbage palms, and *Sabal minor.* In fact, in 1803, the *St. John (Kansas) Weekly News* described Bald Head Island thus: "This island is a bit of Florida anchored off the North Carolina coast."[5] Despite the similarities, there is something pleasantly unfamiliar about the form of development—no lawns and no cars. Residents rely on golf carts and bicycles to negotiate the island, which is only about 3 miles long. Bald Head is believed to be only about two thousand years old and is the southernmost of three islands presumably formed when natural events intermittently conspire to create significant amounts of sand above the tideline. There are lots of cabbage palms on Bald Head Island, many natural, some imported. Bald Head was even called Palmetto Island at one point.[6] The dunes of the Atlantic beach connect all three, but away from the beach the three are separated by tidal marsh. A thinner, older (Middle) island to the north of Bald Head has also been partially developed and is thought to be four thousand years old. And north of that is the most intriguing and best-protected island, Bluff, estimated to be two thousand years older than Middle. Northwest of Bluff Island is tiny Smith Island. The Bald Head Island complex, therefore, is home of the most significant northern and most cold-tolerant ecotype of palmetto.

The Bald Head Island Conservancy exists to study, manage, and interpret the natural areas of the region, and the director, Suzanne Dorsey, allowed me to tag along on a trip to a portion of the maritime forest of Bluff Island. Crossing the marsh from Middle to Bluff on a catwalk was very revealing. Looking down into the marsh that is flooded daily, we could see the trunks of former cabbage palms, hollow husks filled, in some cases, with tidewater. Few phenomena bear witness to rising sea level more than the remains of upland trees deteriorating in a marshy intertidal area.

The ocean-fronting oaks of the Bluff Island maritime hammock are in a restricted area we were unable to visit, but aerial imagery of the preserve shows the palms emergent, protruding triumphantly above the oak canopy. If you look at Bluff Island on Google Earth, you can see the windswept and salt-pruned oaks with cabbage palm canopies protruding far enough above the oaks to cast shadows.[7] These cabbage palms are not the northernmost specimens.

So where is the northernmost naturally occurring cabbage palm? Palm lovers want to know, but our understanding of the exact location keeps shifting due to new discoveries and losses of old populations. The maximum range extension along the coastline was considered by some to be on scrawny Smith Island (just north and east of Bluff Island) at Latitude 33°, 52' and 52.59" North.

But there is at least one conspicuous cabbage palm growing north of Smith Island. And it is the ferry ride from Southport to Bald Head Island that provides a glimpse of a lone silhouette on Battery Island, presumably grown there from a seed dropped by a bird, but possibly washed there and apparently thriving at Latitude 33°, 54' and 40.92" North.

However, about 11 miles farther west there are some palmettos visible from West Beach Drive on Oak Island at Latitude 33°, 54' and 52.63" North. So they are a smidge farther north than Battery Island, and you can see them poking their heads above dense maritime forest vegetation on the north side of the road. Kyle Brown found some farther west in the maritime forest near Shallotte Inlet. If they are still there, they may be slightly farther north. But Brown was looking more than forty years ago, and between planted pine and housing developments, there isn't much maritime forest left in the area.[8] These maritime forest palms challenge Brown's favored hypothesis that the seeds are spread by water, but big storms could conceivably wash the palm fruit into those forest locations. There may not be any naturally occurring cabbage palms north of Oak Island, but there are plenty of unnaturally occurring cabbage palms farther north.

Several years ago I stopped at a roadside nursery in coastal South Carolina to inquire about conspicuous racks of palmettos I'd seen from the highway. The modest rootballs were wrapped in black plastic and the trunks were leaning against a stout timber framework. This display enabled homeowners to pick a particular tree they favored, a level of specificity that is apparently more customary in the Carolinas than in Florida, where cabbage palms are more typically ordered generically, by the linear foot, sight unseen. During the conversation I was told there was a gentleman in North Carolina who was growing cabbage palms commercially from seed.

I don't think I actually laughed, but I was shocked, surprised, and puzzled. Why would anyone take years to grow cabbage palms from seed in chilly North Carolina when they could simply drive down to Florida with a truck and buy all the mature trees they wanted?

It turned out there were several reasons. First, the transportation costs are daunting, and, combined with the cost of the trees, it makes it hard to chauffeur trees northward for hundreds of miles, cover costs, and make a profit. Second, those are Florida trees, not acclimated to North Carolina and unlikely to perform well. Finally, there's an alternative: those naturally occurring North Carolinian cabbage palms. Bald Head Island lies in USDA Zone 8a, which is supposed to have low temperatures between 10 and 15°F.

It stands to reason that plants that reproduce successfully on their own in areas that get down to 10°F are more likely to produce offspring that can deal with cold than plants from florid regions with names like Frostproof, Winter Haven, and Lake June in Winter.

Many plants, including cabbage palms, are cold-sensitive. "Palms have no physiological mechanisms for dormancy so most species are restricted to tropical parts of the globe."[9] If the water inside plant cells freezes, it expands, rupturing the cells and leading to cell death. (Conifers have a variety of strategies, including stronger cell walls and storing water between cells, that enable them to withstand freezing temperatures.) As a result, the USDA and other groups create (and keep adjusting) plant zone maps. The majority of these maps are not based on how much it rains, or how hot it gets, but rather the average annual minimum winter temperature. And palm enthusiasts turn to these maps to gauge where different species of palms might survive.

But average temperatures don't kill plants any more than the average temperature in your kitchen allows you to cook food or serve ice cream. It is the extreme events that are determinant, and there are multiple types of cold. For some palms, it doesn't even have to be very cold to cause injury. An atypical drop in temperature from 70°F to 45°F can kill tissue on some tropical palms. And on clear, calm nights so much heat can radiate upward to the night sky that the leaf surface may freeze, even if the air temperature is above 32°F. These types of cold don't pose much threat to cabbage palms, but hard freezes can be devastating.[10] Duration matters—it can be extremely cold for a brief period, or pretty cold for a long period. Precipitation is another factor—a dry cold is different than freezing rain or an ice storm. And heavy, cloying snow can also be a problem for palms—weighing down and breaking the fronds. So all cold is not alike, and seemingly minor variations in exposure, mulch, plant health, and other factors can make a big difference in whether and where a cabbage palmetto can survive a cold spell.

The palm grower in North Carolina turned out to be Gary Hollar. Hollar lives outside New Bern, about 40 miles from the coast and a degree of latitude farther north than the northernmost naturally occurring cabbage palms. He collected or acquired seeds from that northernmost natural population of Bald Head Island and then went into production. He's developed an elaborate sequence for getting his cabbage palms to market—moving them sequentially from rose pots to one-gallon, three-gallon, five-gallon, seven-gallon, and fifteen-gallon pots in turn. As they grow, these palms have to be given more root space or they will simply bust out the sides of the pots. Gary says he can get them to a seven-gallon pot in four or five years. A seven-gallon pot is a foot tall and about 16 inches across, so a palm in a 7-gallon pot isn't very big, but Gary believes transplanting palms when they are young increases the likelihood they will establish well in the area and consequently be more resistant to cold.

Gary has a number of established cabbage palms at his nursery that he has grown from seed, but he is painfully aware of historic bad freezes in '52, '62, and '77 and the "four below" of 1989, when he lost all his recently transplanted Florida palmettos. Four below is significantly below the point at which leaf damage occurs, which he places around 10°F. His next threshold is 0°F, and if he believes it is going close to 0°F he drags out his secret weapon: the truly old-fashioned energy-wasting incandescent C9 Christmas tree light strings with those paintbox-colored bulbs. Deployed in the

Once a palm develops a trunk, it grows more rapidly. Gary Hollar photographed his daughter next to the same palmetto in 2004 (*top*) and again twelve years later in 2016 (*bottom*). He believes the seed germinated around 1990. Photo by Gary Hollar.

bud area in the canopy, their warm glow protects the bud from freezing. The smaller lights and the newer LED bulbs are useless for keeping plants warm. All his palmettos were damaged in the 6°F of 1996, but they all recovered. Despite freezing weather angst and some coddling, Gary maintains that "I have never lost a palmetto that I grew from seed that grew up in place." And his palmettos made it through the so-called "polar vortex" winters of 2014 and 2015. Then came the prolonged freeze of January 2018. Raleigh recorded six and a half days at or below freezing.[11] At Gary Hollar's it went down to 0°F. His palms were severely damaged, and some were killed by the longest duration of freezing temperatures in one hundred years. Many Florida transplants were killed as well as some from Bald Head Island seeds, but the surviving Bald Head Island type bounced back fast from total defoliation with as many as ten new leaves by August.

In the spring of 2018, the International Palm Society's "Palm Talk" forum was abuzz about reports of five cabbage palms spotted off Fort Fisher Boulevard in Kure Beach, about 3.5 miles north of Battery Island.[12] They are growing wild, but according to Hollar they may well be offspring of cultivated palms he and others provided the Fort Fisher Aquarium.

But New Bern and Fort Fisher aren't the end of the line for transplanted cabbage palms. Virginia Beach is in USDA zone 7b—that's the range next to 8a and it features minimum temperatures of 5–10°F. In October their beachfront cabbage palms look pretty good along the boardwalk, lending an air of tropical sultriness to a location closer to Maine than Florida. In the winter many are cocooned in pastel-colored polyethylene. Nurserymen have told me that north of Virginia Beach cabbage palms are planted as annuals—replaced each spring to lend retailers exotic charm only to die and be replaced, sacrificial emblems of tropicality.

Zone 7b extends up through coastal New Jersey. Does that mean cabbage palms can grow in the Garden State? The answer is obvious. With enough freeze protection, one could grow cabbage palms in Canada. But how far north can *Sabal* palms grow without augmented freeze protection?

The answer, surprisingly, is north of Latitude 47°, which on the east coast is somewhere north of Prince Edward Island. Cabbage palms don't actually grow on Prince Edward Island, but they do at that latitude on our continent's west coast. The explanation lies in the fact that plant hardiness zones aren't based on latitude and that due to warming currents the west coast of the United States is considerably warmer than comparable latitudes on the east coast. As a result, I've seen cabbage palms growing happily in suburban Seattle, which is Zone 8b, boasting an average minimum temperature of only 15–20°F. I don't know how many plants can grow in both Seattle and Cuba, but that's a spread of 25 degrees of latitude, and I'm willing to be impressed by any plant that makes that list.

Nevertheless, I have seen *Sabal* palms growing happily in New Jersey without coddling. Not cabbage palms, but *Sabals* nonetheless, the dwarf palmetto, *Sabal minor*. Joe Kiefer in Franklinville, New Jersey, is another nurseryman and plant boundary pusher, and since dwarf palmetto typically doesn't have a trunk, its belowground bud is better protected from some colds. *Sabal minor* is naturally found as far north as parts of Arkansas and Oklahoma, so it represents the northern vanguard of the *Sabals*. Joe has them growing both at the nursery and at his place, and it is a little geographically confusing to walk down to a creek in New Jersey and see palms waving at you.

The statement that the chair of the Botany Department at Miami has been planting palms around campus doesn't seem too noteworthy until one realizes that the school where he teaches is 800 miles closer to Lake Michigan than Biscayne Bay—so not the University of Miami, but Miami University in Oxford, Ohio. The professor is David A. Francko, and he is part of a band of horticultural boundary pushers, adventurers who are not searching in old places for new plants but rather searching for new places for old plants.

According to Dr. Francko's book, *Palms Won't Grow Here and Other Myths,* there are many species of palms more cold-tolerant than cabbage palms. *Trachycarpus* (windmill palms) can naturally be found above 6,000 feet elevation in Asia and "regularly receive heavy snow" where "temperatures drop well below freezing for extended periods of time."[13] They are the hardiest palms with trunks and can be found growing in Vancouver, southwestern England, and Scotland.[14]

Francko, Hollar, and Kiefer are all part of a loose network of palm aficionados that happen to live outside the Sunbelt. They obsessively gather seeds from palms that have been proven to withstand low temperatures and then plant them in new locations—testing the palms' limits, playing chicken with rogue freezes. Their successes speak for themselves and appear on the internet. The failures rot away in brown-leaved obscurity. Their experimentation shows the most promise for explaining what governs the historic distribution of palms up the east coast. The answer lies somewhere at the junction of seed sources, their means of seed dispersal, and their ability to tolerate cold.

14

———— • ————

Coastal Creep

Natural Distribution, Gulf Coast

The somewhat embarrassing truth is that no one really knows what governed the historic range of cabbage palms. In 1978, the chief dendrologist of the USDA Forest Service, Elbert L. Little Jr., authored volume 5 of the *Atlas of United States Trees*. The rest of the country was covered in four volumes, but Florida required its own separate volume "because it has more native tree species than any other State (except Hawaii), and because it has a large number of tropical species found in no other State."[1]

Little includes the cabbage palm (dealing another blow to those who argue *Sabal palmettos* aren't trees), and the map is startling. According to Little, nineteen Florida counties historically had no cabbage palms, and another five had cabbage palms only in tiny portions of the county.[2] That means more than a third of all Florida counties originally had none or almost none. Little didn't attempt to explain the natural distribution of cabbage palms, but his 256 maps allow one to compare the curious cabbage palm map with all other Florida trees. There's no other tree species that dominates the Florida peninsula, heads boldly up the east coast (out of view), and from Taylor County westward is found in a relatively narrow (less than 10 miles wide) coast-hugging strip that peters out around St. Andrews Bay.

Consider this: starting at the Florida/Georgia state line, cabbage palms extend another 300 miles up the east coast past Florida, but from Levy County west, only coastal counties in Florida are shown as part of the range, and cabbage palms stop altogether around St. Andrews Bay

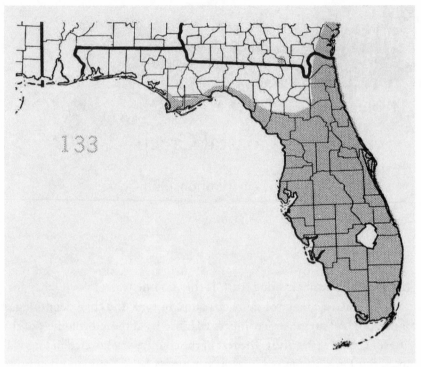

Elbert Little's map (#133) of the latest-known (1978) range of cabbage palms in Florida. *Atlas of United States Trees,* vol. 5: *Florida.*

(Panama City), about 90 miles shy of the Alabama state line. It is perplexing that cabbage palms apparently weren't found in the wild farther west than 100 miles from Tallahassee—roughly halfway to the state line. Why?

The ranges of plants are governed by two things: whether conditions in a given location are suitable to support a plant to reproductive age, and whether some part of a plant capable of starting a new plant (often a seed, but more generally a propagule) can get there.

Well, it seems hard to believe that it's the cold—surely the Panhandle coast of Florida, the so-called Redneck Riviera, is warmer than the coast of North Carolina. Other palms, *Sabal minor* and the other palmetto, saw palmetto, have ranges that extend far to the west. And cabbage palms have been planted all along the Gulf coast and do well. For example, there are plenty of healthy cabbage palms in Baton Rouge and New Orleans.

Turning to the question of Florida's Panhandle, Brown noted that cabbage palms were "seldom seen in nature more than five or ten miles from the sea" and that "the natural range becomes somewhat less than a true continuous distribution west of Apalachee Bay, probably for the same reason as it does in northern South Carolina."[3] Although we infer from Little that St. Andrews Bay (Panama City area) was some sort of natural barrier, both Brown and I have found old cabbage palms on the west side of St. Andrews Bay—Brown on the St. Andrews State Park barrier island, and I on the peninsula to the north.

Brown went on to note plants "established and beginning to reproduce west of St. Andrew Bay," calling them "the vanguard of a slowly expanding range on the Gulf Coast."[4] He expected slow natural dispersal by water, although it is clear now that transplanted landscape palms have dominated the western range extension. Brown believed that long-distance distribution of cabbage palm seeds was primarily accomplished by water currents. Although the seeds sink, the fruit enclosing the seeds floats. And while he acknowledged that small mammals and birds eat the fruit and leave the hard seeds nearby, he was skeptical of the theory that birds carried the seeds long distances. He reasoned: "In the fall when fruits are ripe, succulent, and assumed most attractive to birds, migratory movements are to the south. In the spring when birds are moving north, palmetto fruits are dry, seeds bone-hard, and most have been dispersed from the trees."[5]

To test his theory of water-borne dispersal, Brown soaked palmetto seeds in a saltwater solution. After a month and a half, 90 percent of all the seeds tested were viable and two-thirds produced normal seedlings. Brown figured currents along the east coast could account for the northward Atlantic distribution while conceding "currents along the Gulf coast are not as clearly developed or persistent as along the Atlantic." Still, he thought saltwater dispersal was a feasible explanation of the natural range.[6] We can infer that Brown might have argued that the lack of bays west of St. Andrews Bay was analogous to the lack of bays in northern South Carolina.

But Kyle Brown also acknowledged the role that cold weather plays in determining where cabbage palms can get established: "It seems obvious that low winter temperatures would become the limiting factor not much farther north than the present horticultural range." He conducted experiments showing that low temperatures could affect germination and seedlings and recognized that "ice storms and freezing winds could be a limiting factor in the survival of mature trees."[7] So to summarize Kyle Brown's work from the 1970s, the westward extent of cabbage palms was probably governed by seeds not having an easy time getting there (via currents or wildlife), and the conditions for germination and establishment (lack of islands with bay sides) weren't good. He acknowledged that cold weather could be a factor but appeared to consider it more relevant on the east coast, not mentioning it on the Gulf.

Meanwhile, I was following a botanical group on Facebook when the subject turned to the question of whether native plants can be invasive. Feeling bold with this group of observant and engaged plant profession-als, I floated the following: "I'm eagerly seeking explanations/theories as to why 'originally' cabbage palms were found as far north as Bald Head Island in North Carolina, but only as far west as the St. Andrews Bay area (Pan-ama City) along the Gulf Coast." Brainstormed suggestions were based on a lack of suitable habitat and reasons included low overwashed islands lacking upland ecotones, and low proportions of shell fragments.[8]

If cabbage palms have a weakness, it may be their love of calcareous, limey, alkaline, or so-called sweet soils. They can grow in many places but are not predominate in places characterized by pure silica sand, such as Florida's scrubby Central Ridge. The U.S. Forest Service concedes that "it can tolerate a broad range of soil conditions" but goes on to clarify that *Sabal palmetto* "prefers neutral to alkaline soils characterized by near-surface or exposed calcareous sands, marls, or limestone" and "All soils appear to have one characteristic in common, a high calcium content."[9]

Dr. Scott Zona has studied the *Sabal* palm genus more than anyone else. In 1990, he wrote an eighty-three-page monograph on *Sabal* palms and concluded there are fifteen species in the genus *Sabal*.[10] He not only looked at *Sabals* in the United States but also in Mexico, Panama, Cuba, Bermuda, the Dominican Republic, Trinidad, and Jamaica. In addition, he looked at more than five hundred herbarium specimens and even theo-rized about the evolution of *Sabal* palms.

He begins by mentioning ancestral *Sabal* and *Sabalites* fossils found in what is now the Soviet Union, Great Britain, Alaska, Vancouver Island, and Japan; as well as sites that are more temperate today. It's not that those *Sabal* ancestors were more cold-tolerant; it's just that the places they were growing were at warmer latitudes before they continentally drifted to their current positions in the Northern Hemisphere. Between conti-nents moving northward and glacier-spawning ice ages, palms were pushed southward toward the equator. Palms in Eurasia ran into chilly east-west mountain ranges and were unable to climb over them and retreat south-ward, but palms in the New World found tolerable conditions in the Southeast, Mexico, and the Antilles. Zona then reports that "the modern distribution of *Sabal* (with the possible exceptions of *S. mauritiiformis* and *S. palmetto*) is very similar to its distribution during the last glaciation."[11]

That tantalizing sentence seemed to foreshadow a new paragraph of his opining about the distribution of *Sabal palmetto* (and *Sabal mauritiiformis*, which is found in isolated populations in eight Central and South American countries as well as Trinidad), but no such explanatory paragraph appears. Zona was willing to describe the "northernmost station" of *Sabal palmetto* as Cape Fear (Bald Head Island), but he neglected to discuss the western extent.

So I wrote to Dr. Zona, figuring that if he had spent any time at all ruminating about this puzzle in the two dozen years since his monograph was published, the odds would be good that a compelling, or at least plausible, explanation would have materialized.

His reply led with, "The quick answer to your question about westward expansion is: nobody knows." He agreed with me that the Panama City western extent may simply have been a snapshot—that the range was still expanding and that is as far as they had gotten when people starting mapping. But he suspected a climatological explanation, rejecting my (and possibly Kyle Brown's) alternate hypothesis that the seeds just weren't getting any farther west.[12]

He may have a point. The current USDA Plant Hardiness maps show two gaps along the Florida Gulf coast, areas where warmer 9a (Average annual extreme minimums of 20–25°F) switch to 8b (average annual extreme minimums of 15–20°F). So the climate argument may hold water, making sense of the counterintuitive reality that parts of the North Carolina coast have warmer low winter temperatures than parts of the Florida Gulf coast. Well played, Gulf Stream, well played.

Still, if seed availability played no role—that is, if cabbage palm seeds were being deposited along the Gulf coast wherever 9a zones occurred—wouldn't cabbage palms have developed a disjunct distribution, skipping the colder areas, and colonizing the warmer areas?

The relative ease with which cabbage palms can be transplanted accidently created an opportunity to determine if it was some limiting environmental factor (such as cold) or the lack of a seed source that accounted for the shape of Little's cabbage palm range map. Mature cabbage palms have been planted all along the Gulf coast in order to boost the perception of tropicality, and, as mature trees, the transplants do well, well enough to produce lots of seeds. If cold or some other environmental factor somehow prevents vulnerable seeds from germinating and becoming established, one would expect the mature palms to be lonely, lacking a nearby second generation because their offspring could not become established. If cold or other factors are not a problem, but historically seed supply had been, then there should be many second-generation cabbage palm seedlings along the Gulf coast, surviving the early years and then becoming adults on their own. This tantalizing inadvertent experiment was begging for a road trip, so just before Christmas 2015 I drove 300 miles north to Tallahassee and made a left.

My first goal was to relocate the westernmost original palms that Kyle Brown and I had managed to identify. Elbert Little's range map is small, about 3.5 inches square, but it is clear that there was no shading on the west side of St. Andrews Bay. The palms Dr. Brown and I located are actually south of Panama City, but on the other, western, side of St. Andrews Bay. Kyle Brown had reported old palms in St. Andrews State Park, so when planning the trip I decided to camp in the barrier island park, without anticipating a cold front that would drop temperatures into the high 30's. I found my campsite in the dark, and the headlights revealed a number of young (that is, short) cabbage palms and even a *Washingtonia* palm growing up under mature pines.

I awoke as it was getting light, and my phone informed me that, with windchill, the 37°F temperature felt like 32°F, which is nippy for a Floridian. I left the tent and drove to the easternmost beach parking lot, across the inlet from Tyndall Air Force Base. There were some young palms in the dunes, but I was looking for old, clean-trunked palms. To the north of the parking lot, away from the Gulf, I could see a number of palms. Several had started growing near the edge of the parking lot, but in the early-morning low-sun angle, one farther away really stood out, glowing alone. It was out in a wet area, presumably on a little rise, but it was big and old and so remote it was unlikely anyone had planted it there. Later, after an hour of fiddling around looking at historic aerial photographs, I convinced myself I could spot that same tree in 1964 and 1953—making it old enough to have been seen by Kyle Brown when he surveyed the area on August 3, 1971. I parked at the Gator Pond and didn't see much but concluded later that, had I taken the trail to the south, I would have seen some old palms. I did spot an old palm across the wetland at the Buttonbush Marsh overlook, and I found another old one farther east, and there may well be more, but I had to wonder, given this species' demonstrated predilection for colonizing new areas, why weren't there more?

After packing up my tent, I tagged up with an old friend, Ed Freeman, who grew up on the Magnolia Beach peninsula north of St Andrews State Park. He showed me a palm in the front yard that he swore was about the same size as when he was a kid, and we walked around the house to the palm that his mother moved in a wheelbarrow in 1968. She didn't know "you can't move small cabbage palms," and her naïveté had paid off—it's now about 20 feet tall. There were lots of small palms coming up around the yard there on Magnolia Drive, but I had come to see another really old and tall palm that my wife and I had contemplated once before on a previous visit to the Panhandle. Like the St. Andrews Park palm, it was growing in an area that was wetter than I expected. Ed's parents had settled in a Florida real estate venture so old that the name was not only descriptive, but some of the original inhabitants were still around. In addition to this old palm, there were magnolias, pignut hickories, live oaks, and some very impressive north Florida slash pines. Ed associates ancient palms in the area with higher "middens" typified by more shelly soils and trees such as magnolias. That didn't fit the conditions of Kyle Brown's palms less than 2.5 miles away on the barrier island, but it was worth noting and very reminiscent of Herman Kurz's classic work describing dune succession in which he characterizes the final stage as a hammock of oaks: "twin live-oak, live-oak, or both, pignut, holly, magnolia, and cabbage-palm replacing the pine."[13]

Before I left, Ed told me about some extensive palm hammock on the waterside of the Pelican Point Golf Course on the western end of Tyndall Air Force base on the other side of the bay, and subsequent recourse to Google Earth confirmed his observation, which made having hundreds of old palms just 3 miles away from the scant number of pioneers to the west all the more puzzling.

So what did it all mean? What was going on with the palms on the west side of St. Andrews Bay that had been there in the Park for more than sixty years, ostensibly producing thousands of seeds each year, but with few

This solitary cabbage palm in St. Andrews State Park is believed to have been present in 1971 when Kyle Brown surveyed the area. Photo by the author.

progeny. Why are they there, how did their precursor seeds get there, and conversely, after years of presumable fruiting why aren't there more? Unfortunate soil? Too cold?

I set off hugging the Gulf shore, traversing beachfront resorts and various subdivisions of Panama City Beach and Laguna Beach, which were typical of most twentieth-century barrier island development. The antithesis of this random coastal development lay to the west—the hyperplanned, new urbanist developments of Rosemary Beach, Alys Beach, Seagrove Beach, Seaside (the development that started it all), and Watercolor Beach.

In the haphazard older communities starting with Panama City Beach, everywhere I found mature transplanted cabbage palms, I also found plenty of established smaller volunteers. I even found a palm coming up in some ostensibly undisturbed pinewoods. It was clear that the palms carted from peninsular Florida had no problem growing, flowering, producing fruit with viable seeds, and that at least some of those seeds were able to germinate and grow into a next generation. What I lacked was any negative evidence—a place where neither the developers nor owners had planted cabbage palms. If there were such a place on the Gulf coast, would there be animal-deposited or water-borne cabbage palm seedlings there?

The fortuitous answer lay just west of Rosemary Beach. The new urbanism aesthetic in Alys Beach purports to combine Bermudan and Guatemalan elements, although visitors could be forgiven for detecting a Mediterranean spin, imparted partially not only by the colonial architecture of Guatemala but also by the heavy reliance on imposing date palms (and not, for some reason, *Sabal bermudana*). As a result, Alys Beach had *chosen not* to landscape with cabbage palms, and I didn't see any seedlings. Of course, a half-hour visit in a 158-acre resort town didn't prove none were there—maybe I didn't spend enough time looking.

Back in the car, I kept heading west, crossing Camp Creek Lake and passing the Watersound development before happening upon Deer Lake State Park. I drove in through pine forest and parked and headed to a 1,100-foot-long elevated boardwalk that crossed a dramatic dune field. The plants were sculpted by wind and salt spray—the magnolia trees appeared to be shrubs. There were some saw palmettos, but no cabbage palms.

I crossed Eastern Lake toward Seagrove Beach, which predates Seaside and consequently is a mix of incoherent beachfront development and new urbanist affectations. I stopped at a vacation rental office at the junction of Highway 30a and Eastern Lake Road. They had landscaped with cabbage palms, and not only were there small seedlings germinating below, but some established palms could be seen across the road, and some were so short it seemed unlikely they were planted.

The next stop was Seaside itself, the revolutionary beachfront resort town conceived by Robert Davis, and DPZ (Andres Duany and Elizabeth Plater-Zyberk). It is appealing in spite of the fact that it was assiduously designed to be appealing. Seaside blends efforts at native plant landscaping with fastidious mulching and weed-free, decomposed granite walkways. Despite the inclination to retain and incorporate local flora, Seaside landscapers had brought in cabbage palms (native to peninsular Florida if not the Panhandle), and I quickly found a charming pink house just east of Tupelo Street with volunteer cabbage palms coming up out front. I assume whoever tends the plants mistook these volunteers for the saw palmettos that had been deliberately planted nearby. Who knows how long they'll last because they threaten to eclipse the view of the Gulf from the second floor.

This is the section of Florida's Panhandle coast that is blessed with about fifteen dune lakes. These are usually freshwater lakes set right behind or among beach dunes. It is said these types of lakes occur in only four other places (Madagascar, Australia, New Zealand, and the northern Pacific coast of the United States),[14] although the outwash ponds adjacent to South Beach on Martha's Vineyard are quite similar, if not as duney. Upon leaving Seaside, one is confronted with Watercolor Beach's highway median defined by cabbage palms before one crosses Western Lake, one of the larger dune lakes. In retrospect, these rare and sublime lakes probably never should have been bridged, but about a third of them have been, and at the western end of the causeway there were small cabbage palms growing. How did those seeds get there? Is someone playing Johnny Cabbageseed?

I was tempted to keep hugging the coast, but it was after lunch, and I had planned to sleep in Ocean Springs, Mississippi, which was more than three hours away, so I diverged from the coast and headed to Interstate 10, which has the chops to handle Mobile Bay. But before I could get to Mobile Bay I had to cross Escambia Bay, and just before the bay I noticed cabbage palms growing along the north side of the Interstate. They appeared to be associated with a boat ramp that provided access to the bay. I resolved to investigate on the way back.

My westernmost destination was Biloxi, Mississippi, a Gulf-front town that didn't get as much post-Katrina press as New Orleans but that nevertheless was devastated and, perhaps perversely, it was that devastation that lured me there. My reasoning was that the hurricane that removed so many Gulf-front mansions would have simultaneously created a lot of new territory for palms to colonize and simultaneously reaffirm cabbage palms' general resistance to hurricane winds. Of course, I had no way of knowing how many palms were lost in the hurricane, but I had been to Biloxi a few years previously and noted there were plenty of persistent mature palms that had been there more than ten years and thus had withstood a 30-foot Katrina storm surge and seventeen hours of hurricane-force winds (and probably Hurricane Camille in 1969 as well). These survivors, I reasoned, could provide seeds for subsequent generations and consequently clarify whether cabbage palms could successfully reproduce in southern Mississippi.

Approaching from the east on 90, the first palms I saw were associated with a marina next to the Golden Nugget casino. I drove through the casino's parking garage and found some recently planted palms with seedlings beneath.

Back on Beach Boulevard I turned right, away from the coast onto Cedar Street, just west of the "Church of the Fisherman," the St. Michael Catholic Church, which, without judgment, clearly looks more like a cupcake than most churches. There I found two massive survivor oaks with green retinues of cabbage palms hugging their trunks. That's when I realized I hadn't counted on one of Biloxi's responses to the hurricane. I had assumed that with so many of the buildings gone, a lot of new territory would have opened up for plants to colonize, and I should be looking at young forests of palms, oaks, and magnolias. But the first plants to colonize a cleared area are frequently referred to as being weedy, and Biloxi's leaders no doubt did not want their ravaged waterfront properties to appear weedy. As a result, code enforcement was apparently quite vigilant and most of the still-vacant lots were neatly mown. But that didn't mean there was no room for cabbage palms to germinate and get established. Since the emergent oak roots formed barriers to the mowers, it was common to see cabbage palms springing up around the oaks. Despite the mowing mandate, young palms were easy to find along Cedar. But where did the seeds come from? There was one possible parent on the north side of the church, but I wasn't seeing a lot of mature palms in the area.

I moved away from the Gulf heading west on Howard passing both young/short and old/tall palms. An old home on the corner of Lee had two established palms in the front yard. Across the street to the east I spotted young ones coming up along a property line. Farther west I pulled into a parking lot near the Second Judicial District Courthouse. There was an established palm in the median and a veritable hedgerow of young palms coming up on the north side of the street.

I dropped back down to Beach Boulevard, and, not having had breakfast, I pulled into the Edgewater Mall, lured by a Raising Canes, a chicken franchise we don't have in Florida. I did a little Christmas shopping in the mall and then emerged to discover a large tree well that had been created to protect a preexisting sprawling live oak. I didn't see any mature cabbage palms nearby, yet inside the well there were several dozen young cabbage palms. How did they get there? Did some fed-up mom inform Junior that she wasn't taking him inside the mall unless he ditched his pocketful of palm seeds? Seems unlikely. Not to discount Katrina, but it probably wasn't waves or raccoons. Birds seem like the most likely vector. Bird species observed to feed on cabbage palm fruits as well as migrate include northern mockingbirds, American robins, yellow-rumped warblers, fish crows, pileated woodpeckers, ring-billed gulls, and blue jays.[15]

The coastal development continues west, broken only by inlets. Biloxi gives way to Mississippi City, to Gulfport, to Long Beach, to Pass Christian, to Bay St. Louis, to Waveland—each with their own stories of Katrina devastation and subsequent efforts at revival. It would be easy to push on to see the cabbage palms of New Orleans and coastal Texas. But it was nearly Christmas, and I had seen enough. I left the mall and turned east, traveling the lane adjacent to the beach. I pulled over a few times to look at palms that had been planted, and some that had volunteered, south of Beach Boulevard, that is, fronting the Gulf of Mexico. I hope all readers are regularly awed by the improbability of mature palms apparently thriving in beach sand. I stopped and looked at some, and there were plenty of both germinated seeds and new fallen fruit on the sand below the trees.

The palms I'd seen along the interstate next to Escambia Bay seemed to affirm what I had been seeing elsewhere. They evidently had been planted to decorate a boat ramp, and now a subsequent generation was in evidence.

So what did my 627-mile one-way road trip establish?

The weather, at least in the late twentieth and early twenty-first century, has not prevented the establishment of cabbage palms along the Gulf coast west of Panama City. It is conceivable that colder historical periods may have thwarted germination and establishment, but that doesn't seem likely to me.

Cabbage palm seeds have no problem growing in Gulf coastal resort communities. That doesn't mean cabbage palms could have made it before people disturbed and altered the soils, drainage, and fire regimes, but since cabbage palms seem capable of thriving in a variety of soils, I think we can assume that had large numbers of seeds arrived, some would have made it to reproductive age.

There seem to be lots of seedlings and small palms in the vicinities where mature palms have been planted. Conversely, I typically did not encounter young palms without the presence of supervising adult palms in the general vicinity. But it is not clear how big "the general vicinity" is.

Gravity may play a significant role in determining the original (and current) range of these palms on the Gulf coast. It seems a lot of fruit isn't consumed by wildlife and simply falls, landing at the foot of the tree. Having said that, something is moving cabbage palm seeds around other than gravity.

The anomalous palms growing in the tree well at Biloxi's Edgewater Mall suggests that birds do consume the fruit of established cabbage palms and move them short distances.

These findings push me in the direction of concluding that, prior to people moving palms about, viable cabbage palm seeds never got any farther west than the Panama City area. If birds are responsible for relocating seeds, they apparently have been disinclined to dine and then promptly travel west any significant distance along the Gulf coast. And if currents play a role on the east coast, they don't seem to be having the same success west of Panama City.

It seems clear that when Elbert Little created his map of the "natural distribution or range" of *Sabal palmetto* he was creating a snapshot of a plant that today has the ability to thrive in a much larger region. Whether unknown historic conditions prevented the distribution and establishment of cabbage palms farther west or whether the trees were slowly expanding on their own and would have eventually colonized coastal Alabama, Mississippi, and Louisiana may never be known. What is clear is that people have moved these adaptable trees into new locations along the Gulf coast, that they are escaping cultivation and invading urban and suburban settings when no one is looking, and they have the theoretical potential to invade native habitats along the Gulf coast where they were never found before.

15

Last Tree Standing

Rising Sea Level on Florida's Big Bend Coast

Consider this paradox: One can stand on a high and rocky cliff facing the ocean (think coast of Maine or California) all day watching giant waves crash below. The tides will come and go, the rocks will dry and wet, but there will never be a moment, an instant when you can point downward and say, "I saw it, I saw the tide coming in." The extreme dynamism of the system obscures the subtle changes. But on a flat coast, it is easy to watch the tide arrive as liquid tendrils creep across the landscape. You can walk alongside it as it moves in. And that is what happens in Florida's Big Bend region.

The Big Bend is a section of Florida's Gulf coast that I have described metaphorically as "the broad shallow sweep that keeps Tampa from sprawling to Tallahassee."[1] Without intending to sound negative, geologists refer to it as "low-energy."[2] It has been traditionally defined not by what was there but by what wasn't there—gorgeous barrier island beaches awaiting condominiums, motels, and swimwear outlets. In fact, it has probably been less studied (and consequently is less understood) precisely because "of its outwardly monotonous appearance, low-energy physical processes and the strong interest in sandy, barrier-island coastlines elsewhere."[3]

There is no official boundary for the Big Bend region. Some people in the Panhandle seem to think it stops at the Suwannee River. For others it is synonymous with the "Nature Coast," an ecotouristic name for one of two sections of Florida's coast not already rimmed with concrete. Researchers prefer to characterize the Big Bend as the drowned karst section of the coast that occurs between the Apalachicola Cuspate Delta and southwest Florida's Central Barrier Coast.[4] But the most culturally relevant definition of the Big Bend region is the coast without barrier islands—the section from North Anclote Key (or the Anclote River) to Ochlocknee Bay, near Alligator Point—roughly 200 miles along the coast.

The flatness of the region persists underwater. As a practical matter I have been a mile offshore in the Big Bend and not had enough water to float a kayak. One implication of this supergradual slope is that if the onshore slope is comparable, then a 1-foot rise in sea level could result in roughly a mile of coastal retreat! The potential for dramatic coastal advance and retreat is a geologic reality of the region.

While there are many distinctive aspects of the Big Bend coast, perhaps the most iconic is a forest of cabbage palms fronting a marsh. Naturalist John Muir was sufficiently impressed by Lime Key (one of the Cedar Keys) to make a sketch featuring cabbage palms in his journal. Though we think of John Muir today as a preservationist, at the time he considered himself a botanist. His selection of the palm-riddled Lime Key as a subject may have been influenced by the high regard in which he held the cabbage palm:

"I caught sight of the first palmetto in a grassy place, standing almost alone. A few magnolias were near it, and bald cypress, but it was not shaded by them. They tell us that plants are perishable, soulless creatures, that only man is immortal, etc.; but this, I think, is something that we know very nearly nothing about. Anyhow, this palm was indescribably impressive and told me grander things than I ever got from (any) human priest."[5]

Southeast of the mouth of the Suwannee River there is an oyster bar called Lone Cabbage Reef. Since "cabbage" in these parts doesn't refer to a vegetable in the Brassica family but to the cabbage palm, some might jump to the conclusion that the submerged reef was named for a lonesome landmark tree that once grew there. I know I jumped to that conclusion. The more likely explanation is that the oyster bar is named for an island close to its southern terminus: Lone Cabbage Island. But that, in turn, begs the question of why there would be a lone cabbage island. Many species of palms, it turns out, are remarkably salt-tolerant—a reality suggested by thousands of cartoons depicting one or more shipwreck victims sharing a tiny desert island with a single palm in the center. Or consider the iconic Caribbean or Pacific island hammock slung between coconut palms on a tropical beach. This shopworn mainstay icon of the travel industry, while

89

{Cabbage Palmetto} my first specimen
near Fernandina

contrived, is not inaccurate—coconut palms can grow in close proximity to the ocean. And one palm, *Nypa fruticans,* is called the mangrove palm and forms extensive stands in the estuaries of the western Pacific.[6]

So it is not entirely surprising that the cabbage palm, *Sabal palmetto,* can be found on beaches and next to salt marshes. They can be found surviving closer to salt water than any other native tree, save the mangroves. True mangroves, more than fifty species of trees in sixteen different plant families, can grow with their roots in seawater.[7] This special ability is offset by the fact that they are seldom found inland. Cabbage palms, however, are found in a wide variety of upland habitats, yet they can also tolerate an impressive amount of flooding, saturated soils, and salt water.

Thus a generalized transect from the shallow waters of the Big Bend inland would show offshore seagrasses yielding to salt marsh and scattered, dwarfed mangroves and then to forests dominated by cabbage palms, southern red cedar, live oak, sugarberry, and loblolly pine.

But the palm canopies have shrunk and fizzled, ultimately leaving arrays of what looked like softened phone poles adjacent to the marsh. Something has been killing magnificent mature cabbage palms, palms renowned as true survivors that could withstand fire, frost, hurricanes, and flood. Analysis of more than twenty native trees' susceptibility to hurricane damage has found cabbage palms to be the most resistant to breakage and salt damage and second-most-resistant in terms of uprooting and insect and disease damage.[8] Cabbage palms are found in close proximity to the Gulf of Mexico from Apalachicola (some say Panama City) eastward,[9] through the Big Bend area and all the way south to the Everglades. Three of the habitats found around the Gulf of Mexico (bottomland hardwood forests, sandy ridges, and coastal hardwood hammock on shallow limestone) feature cabbage palms,[10] so the cabbage palm is a conspicuous barometer of upland forest persistence—the last tree standing in many cases.

Actually, using mature cabbage palms as barometers is somewhat misleading. Mature canopy trees are the most likely to persist, whereas more sensitive species or juvenile or otherwise weakened individuals are more likely to respond (and fail) first. This commonsense observation is a reformulation of the observation that when it comes to rising sea level, "tree regeneration fails several decades before canopy trees die."[11] As a result, landscape enthusiasts who were taken with the cabbage palms fronting the marsh were guilty of being seduced by the grandeur of the mature

Dying cabbage palm forest on Florida's Big Bend coast. Photo by Tom Turner.

trees—grandeur accentuated by the fact that the boldly exposed trunks were not eclipsed by the canopies of smaller, younger trees. But this striking phenomenon, tall healthy mature trees, with no next generation, documents the inexorable advance of the sea.

Without a next generation, the current generation will be the last, regardless of whether sea level rise increases, or ceases. As Kimberlyn Williams and her associates put it: "If coastal stands are relict (i.e., non-regenerating), then the causes of canopy death at the coastal margin are relatively unimportant: these stands are effectively dead already. If canopy death lags regeneration failure, then aerial monitoring of coastal forest may fail to detect all but the very last stages of forest decline."[12] So the quest to understand what is happening to coastal forests in the Big Bend became a quest to understand what has turned off the regeneration or recruitment of trees such as cabbage palms.

In the Waccasassa Bay area, something apparently happened about the time parties were working on the Treaty of Versailles and the Grand Canyon was made a national park. No one is sure exactly what happened, but researchers believe some tipping point was reached, and there has been no significant successful cabbage palm recruitment in that area after that (about 1919).[13]

It makes sense that a mature established tree would be better able to withstand stress than a young seedling. But does it make sense that established trees would persist for eight decades while no new trees could take hold? Part of the answer may lie in the challenges cabbage palms face in becoming established trees. As noted elsewhere, the establishment phase can take years and is poorly documented.

Despite the widespread occurrence of *Sabal palmetto* throughout the southeastern United States and parts of the Caribbean, despite its distinction as the state tree of Florida and South Carolina, and despite the significant commercial use of the species (both past and present), little is apparently known about the duration of its "trunkless" establishment phase.[14]

Does the long, and possibly vulnerable, establishment period explain why old palms persist while new palms have trouble getting started in the hammocks facing the marshes of the Big Bend? And how long has this been going on?

Naturalists have long seen evidence of coastal retreat in the Southeast. For instance, in 1849 geologist Charles Lyell correctly interpreted cypress and pine stumps in Georgia's salt marshes as evidence of "a change in the relative level of land and sea."[15] Nor is it hard to find apparently anomalous tree stumps on beaches and in bays—additional evidence that what was once upland habitat somehow became part of the marine or estuarine ecosystem. In 1974, Alexander and Crook found cabbage palm stumps "throughout mangrove fringe forest and on inland mangrove dominated tree islands."[16] But seeing past evidence of a phenomenon is not the same as watching the process unfold in real time, and some of the evidence could possibly be attributed to erosion, storm-related retreat of barrier islands, or subsidence.

Starting in 1939 and again in 1951, scientists noticed dying cabbage palms adjacent to the marshes of the Big Bend.[17] These tantalizing early observations foreshadowed the current awareness of the phenomenon.

Rising sea level can kill upland plant species in several ways.[18] Coastal erosion is obvious—a tree cannot long persist with its roots completely washed out from under it, as happens at Egmont Key, Florida. And while there are scattered areas in the Big Bend where modest bluffs are eroding, this is generally not the case. Changes in hydroperiod, or the number of days the trees' roots are inundated each year, can also kill trees. Thus even along freshwater spring runs, increased sea level could damage trees by increasing the hydroperiod beyond what that species can tolerate. That's because increased tidal heights can exert an effect on freshwater—backing freshwater up and increasing the hydroperiod adjacent to the stream. In addition, salt, expressed either as soil salinity or salt spray, can prove debilitating and fatal to plants. It is also possible that some combination of factors pushes individual species over a threshold or tipping point. Maybe stressed trees are more susceptible to disease or organisms that attack them or they face increasing difficulty competing with salt marsh vegetation.

Researchers from the University of Florida studied upland forests adjacent to salt marsh in the Waccasassa Bay area in the 1990s.[19] This forest was dominated by cabbage palm, southern red cedar, live oak, and sugarberry (*Celtis laevigata*). Cabbage palms were the most widely distributed tree species in the study area. They concluded (based on the order of elimination from forest stands) that relative salt tolerance of seedlings is the best explanation for the failure of recruitment. Tidal flooding may limit replacement.

When trying to understand how rising sea level eliminates coastal forests, it is vexing to tease apart the role of chronic effects like daily tides or gradual increases in sea level from acute catastrophic events such as hurricanes, which can simultaneously impart wind damage, inundation, sedimentation, and the intrusion of salt water far inland (although rains and freshwater runoff associated with most hurricanes no doubt moderate the salt effect).

Even if there were no storms pushing salt water farther into the marshes and up into forests, tidal variation may play a role. Monthly tides are not identical. In fact, an entire sequence of tides that ostensibly brings one back to the same tidal ranges takes nineteen years and eleven days (235 lunar months). This is referred to as the metonic cycle or a tidal epoch.[20] At some points in this cycle, tides run generally higher and at other times, generally lower. It is not known how these years with slightly above- or below-average tides affect the calculus of coastal retreat. So rather than assuming the process is a continuous gradual sequence, researchers now suspect there may be no dramatic changes until an intrinsic threshold is reached, which triggers a noticeable response.[21]

In 2006, some of the leading researchers who have been documenting and unraveling the effects of sea level rise attended the annual meeting of the Ecological Society of America. They brought with them a poster of their most recent findings. A portion of their poster's summary (abstract) reads as follows:

> Forest stands composed solely of adult palms are indicative of the final stage of forest decline before conversion to salt marsh. Recently there have been substantial losses of palms in stands that are at lower elevation and experience greatest frequency of tidal flooding. The mortality of adult palms has accelerated since 2000 compared to observations in the 1990s. The conversion of coastal forest to salt marsh is occurring more rapidly than anticipated from prior trends. The reason for the accelerated losses of adult palms is unknown and requires further study. Hypotheses include increased rate of sea level rise, influences of the regional topography and interacting stressors such as storms and freshwater withdrawal.[22]

For at least two-thirds of a century the dramatic and seemingly plaintive dying palms of the Big Bend coast have attracted the attention of scientists, natural historians, and artists alike. We now know the lonely, barren, and decapitated trunks that drew their eyes were like a magician's misdirection. The "story" wasn't that old mature trees eventually die but that there are no significant numbers of young trees to take their place. As the debate about global warming and sea level rise continues, we can expect more eyes to be drawn to Florida's Big Bend coast, where seemingly minor and subtle differences are wreaking major changes. And when the story of the Big Bend coast is told, cabbage palms will be telling the tale.

Part III

16

—— •——

Hearts of Palm

The Tree You Can Eat

WALT (Startled): Where did you come from?

STRANGER (Mischievously gentle): I ask the questions around here.
(*With his machete, he lops off a piece of cabbage palm and chews on it*)
Swamp cabbage. Keeps yer belly healthy. (*Tossing a chunk to WALT*)
Here, have a chaw.

WALT: Thanks. You got a place around here?

STRANGER: I got a thousand places around here. Every spot of high
ground 'tween here 'n Cape Sable. I was born 'n here and when I'm
three score and ten, I'll die here with the seeds of cabbage palm in
my guts so's a tree'll grow out 'n stand on top o' me.[1]

This exchange from *Wind Across the Everglades,* a 1958 film inspired by
Goodland resident Bud Kirk and written by Budd Schulberg, may have
been many Americans' introduction to the concept that one can eat a tree.

There are a lot of different ways to consume parts of palms, which, for
trees, supply a surprising amount of the planet's food and cover an estimated 100 million acres. Date palms cover 3.2 million acres (2017),[2] coconut palms, 30.4 million acres (2017),[3] and, according to Rainforest Rescue,
oil palm plantations cover 66 million acres.[4] The sap of cabbage palm's
islandic cousin *Sabal bermudana* was tapped to make an alcoholic drink
called Bibby.[5]

Many palm fruits have come into cultivation or are wild-harvested.
And, to be fair, cabbage palm fruits can be consumed, although they are

not very good. They are small, and the tooth-cracking seed inside takes up most of the volume inside the black fruit. Mark Catesby, writing in 1762, found the fruit to be more than palatable: "The berries are globular, as big as cherries, of a luscious sweet taste; and is a great part of the food of the maritime Indians." Catesby is not always reliable, having earlier reported palms growing in New England.[6] And cabbage palm fruits are neither the size of cherries, nor of a luscious sweet taste—Catesby may have been confusing cabbage palm fruits with those of *Sabal bermudana,* which, according to Hodges, have "black, cherry-sized berries."[7]

Personally, I find the taste of cabbage palm fruit to be reminiscent of dates, as is the general experience, since there is a papery outer layer, a vaguely sweet and somewhat granular flesh, and a hard seed. Green Deane's Eat the Weeds website describes the fruit as having a "bittersweet thin fruit coating on the seeds, which are about the size of a pea. The layer of fruit is extremely thin, a skin really, but it does have a prune-like flavor."[8] Reid Tillery, writing in *Surviving in the Florida Wilds,* reports Seminoles would remove stalks of ripe palmetto fruit and then grind both the "rock hard" seeds and pulp with a mortar before serving "on a plate." Tillery agreed "the berries taste vaguely like prunes and are sufficiently palatable."[9] Whether they taste more like dates or prunes, do not be dissuaded from trying cabbage palm fruit by the negative palmetto fruit review penned by shipwreck survivor Jonathan Dickinson and published in 1699. In addition to opining that they had an irksome taste that could take your breath away, he famously described palmetto berries as follows: "They gave us some of their Berries to Eat: We tasted them, but not one among us could suffer them to stay in our Mouths, for could compare the taste of them to nothing else, but rotten Cheese steep'd in Tobacco."[10] So three centuries later about all anyone knows about palmetto fruit is that they taste like rotten cheese steeped in tobacco. But take heart. Here is his description of palmetto: "The Wilderness Country looked very dismal, having no Trees, but only Sand-Hills, covering with shrubby *Palmetto,* the stalks of which were prickly, that there was no walking amongst them."[11] No trees. Prickly stalks. Shrubs one can't walk through. He was eating saw palmetto (*Serenoa repens*) fruit, not cabbage palm fruit.

Coconuts and dates are the best-known palm fruits, but there are others: peach palm (*Bactris gasipaes*), jelly palm *(Butia odorata)*, sugar palms (*Arenga pinnata* and others), and the trendiest is the açai fruit, *Euterpe oleracea,* known for its cedilla that renders the English pronunciation ah-sah-EE, not a-kai. Açai fruits are linked to the consumption of cabbage palm hearts in an indirect way. Because harvesting cabbage from a cabbage palm kills the tree, and because cabbage palms grow so slowly, meeting any significant demand for hearts of palm could have decimated Florida's cabbage palms. But here's where the açai comes in. It is one of a number of palms harvested for their heart, but açai palm has multiple trunks, so careful harvesting can leave the plant alive and capable of further growth. Thus açai became a sustainable source of palm hearts. It was sustainable until 2000, when entrepreneur Ryan Black began marketing the fruit in the United States.[12] Oprah covered it, and the fruit started being promoted as a superfood or superfruit. Açai berries are apparently very high in antioxidants, but the Mayo Clinic is cautious, if not skeptical, noting that açai has been recommended for "arthritis, weight loss, high cholesterol, erectile dysfunction, skin appearance, detoxification and general health," but that "research on açai berries is limited, and claims about the health benefits of açai haven't been proved."[13] In any event, açai palms became more valuable alive as fruit producers than dead as hearts of palm. Now its cousin *Euterpe edulis* is the target of heart seekers—unfortunately it has a single trunk. But we can thank these tropical palms (and the fact that heart of palm is more of a crunchy texture than a taste sensation) for the modest harvest of cabbage palms for hearts of palm (aka swamp cabbage).

One reason vegetarians may feel more comfortable about taking the life of something from the plant kingdom is that, for tree crops at least, harvest typically does not involve killing the plant. Orange trees don't have to die for your glass of juice. Many palms have edible buds, usually referred to as hearts or cabbage. Because cabbage palms have to be killed for their edible bud, it is possible that the term "millionaire's salad" reflected the gratuitous luxury of killing a tree for a crunchy morsel that could be approximated with easily grown water chestnuts.

John Bartram (William's father), 1765: "Here we cut down three tall palm or cabbage-trees, cut out the top bud, the white tender part or rudiments of the great leaves, which will be six or seven feet long when full grown, and the palmed part 4 feet in diameter; this tender part will be three or 4 inches in diameter, tapering near a foot, and cuts as white and tender as a turnip; this they slice into a pot and stew with water, then, when almost tender they pour some bear's oil into it and stew it a little longer, when it eats pleasant and more mild than cabbage: I never eat half so much cabbage at a time, and it agreed the best with me of any sauce I ever eat, either alone or with meat. Our hunters frequently eat it raw and will live upon it several days."[14]

In 1822, "a Recent Traveler in the Province" reporting on the Seminole Nation of East Florida, noted: "The orange groves are a great support to the Indians, who generally roast the orange previous to eating it. With this fruit, the palmetto cabbage, and the wild potatoe, they are often enabled to live, without either hunting or cultivation. They frequently encamp for months along the river, sometimes in the deepest swamps, subsisting on these *wild provisions,* as they usually term them. Their being able to exist in this way, will prove a great obstacle in any attempt to wean them from their erratic habits."[15] Such accounts suggest precolonial life in Florida, but not only were oranges brought by the Spanish, but the Native Americans in Florida prior to the mid-eighteenth century weren't known as Seminoles.[16] So despite swamp cabbage being part of the Seminole diet, it is believed that swamp cabbage has only been eaten for less than five hundred years (coinciding with the arrival of metal axes in the New World) since "the utensils available to the indigenous people were not adequate to extract the heart efficiently."[17] In other words, without a functional axe it would take more energy to harvest the heart than could be gained by eating it. On the other hand, an old-timer told me of a burly man he knew who would seek out a fallen, but living, cabbage palm growing prostrate close to the ground, grab the canopy leaves in both hands, lift the palm head up and then whack it down on the ground in a manner that allegedly separated the heart from the trunk, which is probably how Paul Bunyan would have done it if he hadn't been so heat averse.

In 1859 the botanist F. Andre Michaux started his third volume of his *North American Sylva* with a description of the cabbage tree, which included the following: "The base of the undisclosed bundles of leaves is white, compact and tender; it is eaten with oil and vinegar, and resembles the artichoke and cabbage in taste, whence is derived the name of Cabbage-tree. But to destroy a vegetable which has been a century in growing, to obtain three or four ounces of a substance neither richly nutritious nor particularly agreeable to the palate, would be pardonable only in a desert which was destined to remain uninhabited for ages."[18]

Because a normal cabbage palm has only one growing tip, one bud, harvesting *Sabal palmetto* hearts is a distinctively unsustainable endeavor, at least for the tree in question. In places where cabbage palms are uncommon, killing one for a portion of a meal is likely to be viewed as a questionable decision. But on the ranches of Florida's lower peninsula and the forests of the Big Bend, it is hard to see how cutting some cabbage is going to make any appreciable difference.

In 1922 the Bass Grocery and Market in Tampa was selling five pounds of sugar for forty cents, a dozen lemons for a quarter, alligator pears (avocados) for five and ten cents, and palmetto cabbage (known as swamp cabbage) "good size and fresh" for twenty and twenty-five cents.[19]

In 1955 the *Fort Myers News-Press* recognized eighty-two-year-old George Roland as the "Swamp Cabbage King." Born in 1872, Roland harvested and sold cabbage palm "buds" for thirty years, getting fifteen cents for one during the Depression and a half dollar during the Second World War. George believed the best hearts come from small trees, "when the bud is almost as high as a man's head." He used an axe to create a finished product 6 inches thick and 18 inches long. One bud would "make a main dish for two meals for two people."[20]

In 1982, the Junior League of the Palm Beaches published *Heart of the Palms: Favorite Recipes of the Palm Beaches.* The cookbook attempted to feature regional specialties, and the first chapter includes "10 recipes for everything from Hearts of Palm au Gratin to Hearts of Palm Dijon." The article announcing publication incorrectly asserts that cabbage palms are a threatened species and recommends ordering some from a nursery in Okeechobee for $2.50 each. "One heart makes 4 cups, according to the book."[21]

In February 2007, I paid for a booth at the Swamp Cabbage Festival in LaBelle, Florida, and rustic entrepreneur and woodsman Willy the Losen and I hung out a banner that read: "If you have questions about cabbage palms, you need to talk to us. If you have answers, we need to talk to you."

The most common question boiled down to, where's the cabbage? Where indeed! It seems a measurable constituency of the festival-attending public expected some sort of spherical palm-cabbages that hung coconut-like beneath the fronds and that might possibly be picked with a long pole. I tried a number of metaphors before finally realizing the spherical cabbage image was confusing them and that a stalk of celery might hold the key. I reminded them that when they bought celery, the bigger, outermost leaves

George L. Espenlaub, Everglades guide, harvesting palm for swamp cabbage. Courtesy of the State Archives of Florida.

were relatively fibrous and green, and as one moved inward, the smaller, still developing leaves became softer and paler and less fibrous. Then I encouraged them to view the top of the cabbage palm as if it were a massive stalk of celery and to mentally pull off the outermost leaves, moving toward the center, or heart, of the palm. Eventually they would encounter a crunchy, whitish mass of plant tissue cylindrical in shape and in taste not unlike the center of a head of cabbage or artichoke heart. Since palm buds tend to be big, many palm species have similar edible "cabbage" within. More than fifty palm species have been harvested for their edible, relatively undifferentiated meristem plant cells.[22] The hearts of some palms, however, are not merely bitter, but poisonous, so restrain any "hold my beer" inclinations when encountering a random palm heart lying about.

Nowadays, you don't often see freshly harvested swamp cabbage for sale. My wife spotted a trio selling some from the tailgate of a pickup truck off of Highway 98 in Cross City, Florida. Ten dollars was the going price in 2015. (At the end of the Depression they sold for three cents apiece.)[23] They picked one for us that they promised would last a few days and explained how they cut the "fans" off with a machete and then chop the cabbage off with an axe. We were cautioned to never buy cabbage cut with a chainsaw since the palm is living and that saw oil would get sucked up into the cabbage and ruin it.

Because so few people actually attack a cabbage palm with intent to eat its heart, those who do tend to regard their skills as impressive, while questioning the elegance and efficiency of others. My first cabbage harvest was with Leslie Gaines, a cracker with access to a ranch his family used to own. Armed with an axe and a chainsaw, we took off in a Kubota four-wheeler and drove past hundreds of cabbage palms that I thought could be candidates for consumption, and, after more jouncing around in the lumpy pastures, I finally asked Leslie what he was looking for and was surprised by his answer. He did have a preferred height he was looking for, one based on where his axe swing would be most efficient. But Leslie is a nature film director, and he was also looking for a palm whose beheading wouldn't mess up some future wide shot.

We arrived at one he concluded was not likely to make it to DVD, and he held his axe up with the axehead even with a point from which the youngest fronds seemed to emanate. He then related that the bottom of the axe handle indicated where he would be chopping. Before one can start swinging an axe one has to remove the spiky bootjacks, and that was accomplished in about ten minutes with the chainsaw. I scurried about pulling the fallen fronds away from the base of the tree because it is good to minimize tripping hazards when wielding a chainsaw, even if they are downward cuts close to the tree.

Once the tree was shorn of its boots, Leslie went to work with the axe. Attacking a cabbage palm with an axe requires a different strategy than the one used when felling a cypress. Rather than hacking a large notch in one side of the tree and then making a back cut on the opposite side, exposing the palm heart requires circling the tree repeatedly chopping through the bootjack remnants and yanking them off. The process reveals muscular, pale, smooth material below the leaf that morphs into copious russet webbing. This webbing looks remarkably like loosely woven fabric. So much so that it is easy to imagine this material might have been the inspiration for woven fabric since there is a visible warp and woof, but it consists of two overlapping layers and is not actually woven, over-underwise.

Leslie ended up with a cylinder about 4 feet long. The butt end was about the diameter of a football, and it tapered somewhat to the top end, where the fronds had been shorn off. We retreated to a cabin, where the dissection began in earnest. That involved more external leaf removal. The white heart oxidizes quickly, so the chunks that are broken or cut off are quickly thrown in a pot of water. Leslie cooks it in chicken broth to which he adds cooked bacon (it is not easy to find the right salt pork in the grocery), butter, and Everglades Seasoning. Everglades Seasoning is a spice mix that lists spices as the second ingredient, after salt. As the cabbage cooks, a scummy foam forms on top, and he skims that off. I'm of the opinion that almost anything cooked with chicken stock, bacon, butter, and spices is going to be tasty—sort of a stone soup concoction. Consequently, I prefer the crunchy, raw swamp cabbage, which, if you use your imagination, at times almost has a hint of coconut. I like the recipe of Marjorie Kinnan Rawlings, who suggests: "Slice thinly and soak for an hour in ice water. Drain well, serve with French dressing or a tart mayonnaise. The flavor is much like chestnuts."[24]

My experience is that people who cook swamp cabbage, particularly if they cut the cabbage themselves, consider themselves to be Florida crackers. Definitions of Florida crackers vary, but here's mine. A Florida cracker is someone born in Florida who possesses at least one of the following two additional traits: (1) several generations of ancestors who lived in Florida, and/or (2) a strong rural outdoor connection. And every cracker I've met believes their approach to cooking cabbage is not only authentic but discernibly more authentic than most any other cracker's approach. Not being a cracker, I'm not in a position to officiate, so I defer to again another non-cracker author, Marjorie Kinnan Rawlings, who wrote: "Only an expert can cut down a palm and strip the core properly to its ivory-white layered heart. The lower portion of the heart must be tested by taste for bitterness, the upper portion for fibrousness, until one is down to a white cylinder of complete sweetness and tender crispness." Her camp-style recipe called for white bacon and a little pepper. She suggested cooking for forty-five minutes or until tender. For her own table she went with butter and salt cooked in very little water and then simmered in hot cream. "Prepared this way, heart of palm is fit for a king."[25] Conservation writer Ernest Lyons also offered two alternative approaches: "There were two schools of cabbage palm cookery. Either one turned out a product fit for the gods. Most frontier woodsmen preferred to simmer a few dozen diced cubes of salt pork in the bottom of a cast iron Dutch oven, dump in the sliced palm cabbage and semi-fry it, then throw in a cup of water and slow steam it with the lid on tight. Another method was to add a lot more water, shake in a little evaporated milk and a few slices of onions, allowing the pot to simmer for an hour or two. A few regarded the addition of the onions and canned milk as heresy."[26]

Seven years after my booth experience, Leslie joined me on a visit to the fifty-first Swamp Cabbage Festival. Les was hoping to find Buddy Mills, the man with the custom swamp cabbage axe, and I was hoping sawmill operator Ron Warner might be there. I'd also gotten to know the 1979 Swamp Cabbage Queen, so I felt even more connected to the arcane event, which for all I know may be the nation's only festival based on killing and eating trees. In an effort to encourage Les to make the 120-mile drive, I hyperbolically promised all manner of swamp cabbage food products—I might actually have mentioned swamp cabbage Jell-O.

We parked south of the river and walked toward the commotion in Barron Park on sidewalks lined with massive bromeliad-laden oaks and cabbage palms draped in strangler figs. The stage under the oaks looked the same, as did the eclectic mix of bikers and retirees listening to country rock (the Posse Band) featuring a lead singer in a DeSoto County Sheriff polo shirt. The pair of handcuffs on his belt lent a note of enforcement authenticity. The novel, but predictably underwhelming, armadillo races were still there, as were the booths separated by low walls decorated with skirts made of palm fronds. We walked by numerous booths but stopped to chat with representatives of Defenders of Wildlife and the Conservancy of Southwest Florida, who sat in a chain-link cage, which was a model of what the groups would help build to protect your goats, chickens, or whatever livestock from marauding Florida panthers. There were booths pushing local real estate, Seminole Indian beadwork, leather hats and belts, mineral specimens, and slingshots. And every tenth person seemed to be walking around with a four-foot walking stick, which we learned were free at the tent of the Fellowship of Christian Farmers, if you were willing to listen to a one-minute spiel about heaven, sin, the blood of Christ, purity/ forgiveness, and spiritual growth.

But the role of swamp cabbage seemed to have diminished. In fact, if it wasn't titled the Swamp Cabbage Festival, you could be forgiven for assuming it was an armadillo festival or a walking stick festival. Three joints were selling swamp cabbage fritters, and another was selling a small jar of pickled swamp cabbage for five dollars ("You'll want two jars because once the wife has some, you'll need a jar to keep up") as well as cut sections of palm, which seemed short to us, for ten. Leslie looked askance at a vendor who, in Leslie's opinion, was abusively overcooking his cabbage. One booth promised swamp cabbage gumbo. But we never found a swamp cabbage demonstration, there was nothing approaching swamp cabbage Jell-O, and there was no sign of Buddy or Ron. We left town with a jar of unlabeled, pickled swamp cabbage, a jar of mangrove honey, a beaded necklace, and two 4-foot-long walking sticks. We may not have been saved, but we were unlikely to stumble.

I'd first run into Buddy in Tampa, where he demonstrates swamp cabbagery every year in February at the Florida State Fair. His setup under an oak tree is removed from the carnies, pitchmen, and chocolate-covered bacon that dominate the rest of the fair. He's found in a collection of historic buildings that had been scrounged from around the state and resettled on a section of the fairgrounds known as Cracker Country, and Buddy and others are there to convey cracker tradition. While others press cane syrup or cook cornbread and collard greens, Buddy prepares swamp cabbage all day with a running patter as he fields questions from curious fairgoers. Most watch silently, but some are willing to challenge his expertise, which, based on watching him for several years, is usually a mistake because Buddy can use both humor and his hard-won knowledge about the state tree to disarm critics and charm the crowd.

At the state fair in 2015 I told him I wanted to go out with him when he was cutting cabbage, and in mid-June we finally connected. I got up at 5:30 A.M. to drive two and a quarter hours to meet him in Okeechobee. We headed west on State Road 70 past the Griffin Trees palm nursery, crossed what is left of the Kissimmee River (known to water managers of the South Florida Water Management District as C41-A), and turned south on Fulmar Terrace. Buddy told me the Seminole Tribe of Florida owned a lot of the surrounding land. We crossed a big drainage canal (the mellifluous S-84), went through an area known as the cabbage woods, and passed a sod farm, driving until we came to some abandoned pasture where Buddy had been given permission to cut. As was the case with Leslie, Buddy employed an aesthetic sensibility, having been told he could cut cabbage but not where it was real obvious. We bounced around in abandoned pasture until we found some small oak and palm hammocks with lots of small palms coming up on their peripheries.

This landscape had been wet prairie with isolated tree islands of palm and oak. This site is north of Lake Okeechobee, so technically north of the official Everglades, but it behaved like everglades—shallow, wet prairies too wet for palms and oaks with scattered marginally higher hammocks oriented with the overland water flow, like long, linear stretchmarks on the surface of the planet. At some point drainage ditches had been etched into the landscape, and the drier conditions have allowed palms to germinate and get established outside the historic hammock islands.

Buddy introduced me to two helpers: Jimbo and Josh, who had followed us in. There are three ranks of cabbage harvesters. The lowest are toters who tote cabbage back to the trucks or gas cans to the more experienced. High school football players are apparently a good source of toters since the skill set seems to feature strength, a tolerance for heat, and an ability to follow basic directions. More skill is required for the apprentices, who start trimming the fronds in advance of those who know how to use an axe. Josh was an able apprentice who wielded a knife historically used for cutting sugarcane. Advancing to the most skilled level requires knowing how to use an axe properly (some use chainsaws) and how to distinguish bad cabbage from good cabbage. Far more people know how to use an axe or chainsaw than know precisely what part of the tree's core is edible.

Before they started in with axes, cane knives, and chainsaws, passing reference was made to the presence of ground rattlers (pygmy rattlesnakes), perhaps to gauge how much of a city feller I was, but once they could tell I greeted snakes more with respect than fear, they probably assumed I was not a candidate for unexpected fainting or screaming. Funny thing about those so-called ground rattlers, though—they told me it is not unheard of to find them hanging out in the bootjacks of the palms, where they dine on passing tree frogs.

I didn't know it at the time, but on the drive out Buddy had surreptitiously subjected me to a litmus test. He needed to know if I was some "National Geographic" type who was going to use photographs of palm harvest as some sort of evidence against him. He told me palm harvest was illegal in about eight Florida counties, and he wanted to make sure I wasn't part of some palm tree liberation front. I reassured him by opining about how invasive cabbage palms can be.

There isn't much to fear when cutting cabbage, aside from daunting heat, the aforementioned ground rattlers, poison ivy, invasive fire ants, bull ants, scorpions, and the two types of wasps—Guinea (*Polistes exclamans*) and the little paper wasp (*Mischocyttarus mexicanus*)—that nest on the undersides of the palm fronds. While we were working, Jimbo got stung on the arm by a scorpion, and I inadvertently stood on a fire ant mound but somehow noticed my error promptly and escaped.

Cabbage harvest proceeds through a number of stages the whole purpose of which is to convert a small tree into some modest portion of portable food. Josh wielded his cane knife and circled a palm, undertaking the first step of removing the fronds, a process Buddy refers to as "trimmin'." The purpose of trimming is to enable the harvester to get close enough to the trunk to swing an axe. Nowadays trimming is more frequently accomplished with a small chainsaw, which is what Buddy was using that day. I circled one freshly trimmed palm and counted twenty-three green fronds on the ground.

Then Jimbo would approach with his custom double-headed axe. I found he could accomplish the second stage of palm harvest ("ringin'") in seventy-five seconds, using twelve swings of the axe. His axe, like Buddy's, has been custom made or modified to have a much wider, rounded blade that made a much broader cut in the palm boots. Ringing involves cutting slightly downward toward the heart (horizontally when using a chainsaw) through the bootjacks and pseudobark so that the diameter of the trunk is reduced to the area where white developing leaves and trunk predominate. The downward chop angle minimizes the rebound that can occur when striking the resilient bootjacks sideways. In addition to the downward chops, the ringer has to stop periodically to remove the bases of the fronds, which completely encircle the trunk. The result is a vertical column of the center of the palm that is roughly half the diameter of what was started with.

During a break I asked the crew how long they thought cabbage palms live. Jimbo replied "Forever," without hesitation, confirming the views of the old-timer interviewed by Stetson Kennedy, who commented, "You can't see the end of it lessen you cut it down."[27]

Once the ringing is done, a simple chop separates the heart from the trunk, and it is carried to a tailgate. Buddy and his son, Chad, working as a team—with Chad trimming and Buddy ringing—can harvest twenty cabbages an hour, if the palms are close to each other. Buddy told me he can harvest fifty a day working alone and that he cuts two thousand buds a year. Tom Gaskins claimed to have cut 105 in one day and credited Frank Jones with cutting 305 in one day "to feed his hogs."[28] Swamp cabbage is sometimes portrayed as a survival food, but unless you are lost with a chainsaw and axe, your time and energy would be better spent on virtually any other pursuit. In the time Buddy and Chad can harvest twenty, a first-timer might be lucky to have one cabbage and would likely be blistered and exhausted.

When they are done cutting, they converge at the trucks, open coolers of ice water and commence "bootin' 'em out," which involves shucking the tough outer leaves away in search of the white heart. The central core of the palm is relatively easy to cut, tougher than the white part of watermelon rind, but not much. The bottom of the heart is about the diameter of a softball, and Buddy and Jimbo used small knives, cutting toward themselves to dissect what was left of the plant. Jimbo observed that "You don't cut it with a sharp knife," reasoning that a sharp knife makes it too easy to inadvertently cut bitter material into the pot.

The leaf bases are pulled off, and any material that can be snapped cleanly off the bottom goes in the ice water, which keeps the cabbage from

Buddy Mills's special cabbage palm axe has a distinctively curved blade.
Photo by the author.

oxidizing and turning brown. If it doesn't snap cleanly, it is rejected. As leaves are removed, the diameter of the heart keeps decreasing until it becomes possible to just cut across the entire mass of developing leaves. You can overcook swamp cabbage into mush, and you can add too many overwhelming ingredients (cream of mushroom soup!), but the main enemy of good swamp cabbage is bitterness. That's because not all of the white "heart" material is suitable for eating—some of it is quite bitter, and Jimbo and Buddy swore that even a small piece of the bitter material could ruin a potful. Buddy's daddy told him blooming palms are more bitter, and recently he's had experiences that tend to support that observation. Others believe you need to avoid old palms and ones near the water.[29] Buddy is careful to store the buds horizontally or, if stacked leaning, to keep the original orientation of the bud. Storing them upside down apparently allows the bitter components to seep downward and ruin the cabbage.

The pale, developing tissue hidden from sunlight consists of two types of organs—the developing leaves and the developing trunk. The developing leaves near the center of the plant are what you're looking for, but as you move farther away from the center of the plant they get tough. The developing trunk, which appears very similar to the leaves, is bitter. So Buddy and Jimbo were very discriminating. Buddy looks to see that the developing leaves separate from each other, somewhat like onion rings separating, but if it is a unified mass (future trunk) it is rejected as being bitter.

It is hard to overestimate the lurking bitterness. *The Carolina Housewife,* written in 1851 by "A Lady of Charleston," advises: "Trim off carefully the hard folds of the palmetto cabbage, then boil the inner part for two hours, during which period the water must be changed three times, that the bitter quality of the cabbage may be entirely extracted."[30]

They cut twenty-five cabbages that morning, which was part of a much larger order for a benefit dinner Buddy was helping with. On the drive back to Okeechobee I learned that Buddy was a schoolteacher but had been employed for a while in the search for medflies, *Ceratitis capitata,* the larvae of which feed on a number of commercial fruits. Buddy had been seeking out plants such as guavas that could harbor the flies, and, as a result, he can spot a number of fruit trees at 60 mph that the rest of us would have trouble recognizing on a walk.

Florida could use a few more like Buddy Mills, crackers with a lifetime spent in proximity to the natural world who are willing to share lessons with tourists, new residents, fairgoers, and those Floridians whose main experience of cabbage palms is paying a lawn service to overprune them.

17

---•---

Palm Entrepreneurs

Money Growing on Trees

The town of Perry in Florida's Big Bend region is home to the Forest Capital Museum, which "celebrates the timber that built Florida. The heart of the museum is dedicated to long-leaf pines, which grow on the museum grounds, and the 5,000 products manufactured from them."[1] The life-size dioramas cover Florida's pine industry, and there's a display about cypress, but there isn't mention of any Florida forest industry based on cabbage palms. The fact is that most foresters don't spend too much time thinking about cabbage palms. Kendrick and Walsh wrote a 585-page book, *A History of Florida Forests,* affording cabbage palms four mentions.[2] Laurence Walker's *The Southern Forest* devotes a four-sentence paragraph.[3] While it is true that palm forests represent only about one-fiftieth of Florida's forests, as of 2015 it was estimated there were 108.34 million cabbage palms in the state[4]—surely a forest resource of some significance, even if foresters have tended to ignore palms because palms aren't used for lumber or pulp—a case of not seeing the trees for the forest.

Cabbage palms persistently resist becoming lumber. There seem to be several explanations for this: the modest diameter of the trees, their fibrous nature, a complaint that they are too wet, and the fact that, counterintuitively, they have earned a reputation for destroying saws that breeze through pine or cypress. Tom Gaskins, introduced earlier, simply said, "Sawing a cabbage palm log is about as wearing on a saw as sawing a block of cement."[5]

A sawyer, coincidentally named Tom, writing in the online Forestry Forum described his experiences when he tried to convert a palm log into planks: "Cabbage Palm is a Monocot and has no internal structure like you would expect to find in a tree. It is very fibrous and contains something that wears the teeth of the blade almost immediately. I could cut only one trunk before the blade was so dull as to be almost useless. The boards are pretty and remind me of matted grass. Perhaps a wall in Pacific motif would look ok but I wouldn't saw one again unless I arranged special pricing. One blade per trunk is too expensive for the sawyer to foot the bill and only the bottom of the trunk makes a decent board. This tree has a heart that is edible. That's why it is called a cabbage palm. The closer you get to the top of the tree where the cabbage is, the less structure the tree has. The 'boards' are as limber as a noodle and can barely be carried by two men because of the sagging. How they will dry is still to be discovered."[6]

A 1946 photograph in the Florida Memory collection shows two men sawing a cabbage palm log. The caption reads: "When C. L. Baker and his helper, Edgar Huff, reach the step of sawing against the palm's fiber, they arrive at the real problem of building with palm. The palm chewed off 3 inches of the steel blade of the electrically operated saw, and the blade must be resharpened for each new log. According to Baker a cutting press instead of a saw may solve the problem."[7]

Some say the trunks are full of sand, and that's not far from the truth. Quartz beach sand is silica, silicon dioxide, SiO_2. Cabbage palm trunks contain polymerized monosilicic acid $(Si(OH)_4)$. These silica bodies occur in small cells called stegmata. Researchers preparing palm material for microscopic analysis using a microtome (a sharp knife for slicing thin sections) are advised to dissolve the silica bodies first using hydrofluoric acid—that's the acid that has to be stored in plastic bottles because it dissolves glass. So how do the sandy polymerized monosilicic acid stegmata bodies get there? "Its limited appearance in visible form in plants may be the result of incidental absorption in its soluble form by roots, subsequent distribution by the vascular system, and final sequestration in a solid polymerized but opalescent form."[8] So, metaphorically at least, cabbage palm trunks may have more of a future as sandpaper than as toilet paper.

Another Forestry Forum post by "Mark" described an attempt to convert cabbage palms into plywood: "About twenty years ago the Ga. Pa.

C. L. Baker and Edgar Huff sawing a palm log, New Port Richey, Florida, 1946. Courtesy of the State Archives of Florida.

[Georgia Pacific] veneer mill near Palatka, Fla. made some plywood out of cabbage palms. I actually saw a sheet of it. It is possible that was the ONLY sheet made. I had expected something akin to a grass skirt, but was surprised at the appearance. The fiber was tight, and relatively smooth, looked good! The GP employee told me that it shut down every machine as it went through the mill! Don't expect to see any in your box store!"[9]

Since cabbage palms can't be sawn into usable lumber, the proposition that the cabbage palm trunks lack utility seems entirely plausible, because it's not obvious what use they could be put to. Their most mundane use was to construct corduroy roads—stretches of logs laid perpendicular to a roadway to facilitate travel through marshy areas. Palm log corduroy roads were constructed in places like Bispham Road in Sarasota, Bailey's Bluff Road in New Port Richey, and the road to Hickory Island near Yankeetown.

With all due respect to longleaf pine, longleaf is (was mostly, since its range has been reduced by an estimated 97 percent)[10] used in only two basic ways: as lumber, and as a source of products made from the sticky sap:

pine tar, pitch, turpentine, pine resin, and rosin. These gummy siblings in turn yield the other 4,994 derivative products featured in the Forest Capital Museum.

Despite the general disdain foresters show for cabbage palms, there's been a growing forestry interest in, and awareness of, NTFP—Non-Timber Forest Products—which are sometimes sorted into culinary, wood-based, floral and decorative, and medicinal and dietary supplements.[11] Cabbage palms can hit all those bases, having been used as food, to make brushes and whisk brooms, lightning-thwarting boondoggles and Palm Sunday crafts, and a Seminole treatment for "Grass Sickness."[12] In 1901 the American Fiber Company tried making "cathartic tablets" from cabbage palm berries, but this remedy never caught on.[13]

There are many species of plants and animals that exploit cabbage palms, but humans are unique in their ability to combine imagination with manipulation. Like the longleaf pine, cabbage palms have been used structurally for the construction of dock pilings, bridge supports, forts, porch supports, and even homes.

In 1809 the Agricultural Society of South Carolina presented Colonel Shubrick with the Society's gold medal for his experiments using cabbage palm planks for boat sheathing immune to "the attack of those worms which are so injurious to all other kinds of wood under water."[14] In 1827 wharf builders in Pensacola placed an ad for 1,500 cabbage palmetto tree logs, from 12 to 22 feet long.[15] And, pursuant to a message from President Andrew Jackson calling for a Radiating Marine Rail-way for the repair of sloops of war at the Navy Yard at Pensacola, there was a demand for 1,000 palmetto piles and 16,000 board feet of palmetto planks, 4 inches thick and 1 foot wide. Sticks (trunks) 16 to 36 feet were estimated to cost seventy-five cents each, and longer "logs 30 to 40 feet are the largest that could probably be procured would cost about a dollar a piece."[16] For a while cabbage palm logs were a preferred choice for dock pilings in places such as Key West and Cedar Key. They were capped with metal and said to last thirty or more years.[17] As late as 1960 people were advertising cabbage palm pilings for docks at one dollar apiece.[18] But their use as timbers or pilings has been overshadowed by other, fibrous, uses.

Hernando D'Escalante Fontenada recorded men wearing breech-cloths woven of palm and women wearing a shawl made of a kind of palm leaf, split and woven.[19] Fontenada's account from the middle sixteenth century is the first historical account of Florida palm fiber, although archaeological research has documented much earlier palm fiber craft. The cabbage palm can be viewed as an upright cupboard of five fiber types: three visible and two normally hidden: the single-strand hairlike filaments found between the leaf segments, the russet tufts found at the base of the bootjacks, and material between the forks of the petioles that looks like webbing. In addition, there are the long, straight bristle-like fibers found hidden inside the base of the leaf stem (petiole) and wavy fibers in the decomposing trunks that are reminiscent of shredded wheat.

The visible fibers have long been employed as cordage. Observant people could easily have been inspired to experiment by watching birds use the finer filaments to line their nests or seeing squirrels tugging at the russet fibers for their nests. Both the long hairlike filaments between the leaf segments and the reddish palm trunk fiber can be made into usable cordage by a technique called plying that involves simultaneously working fiber bundles, twisting both of them in the same direction, but then twisting them together as a pair in the opposite direction, which prevents them from untwisting. But what it is less clear is how individuals assessed the potential of the unseen fibers, those hidden in the leaf bases.

Before the development of synthetic fibers, it is hard to overestimate the extent to which fiber (frequently spelled "fibre") research was an American industrial preoccupation. At that time, most fabric, and all cordage, twine, rope, and much upholstery was made from plant fiber. Nearly one-fifth of *The Report of the Commissioner of Agriculture* for the year 1879 was devoted to "Vegetable Fibers." The general obsession with fiber is reflected in a 1926 enumeration of fifty potential articles that may be manufactured from fiber, a list that included "carpets, auto top cloth, tire lining, upholstery, cravats, sunshades, underwear, bookbindings, laces, wire insulation, brake linings, hampers, surgical dressings, buntings, paper, etc."[20]

Sometimes the ambitions of cabbage palm boosters got the better of them, or their investors. In 1951 an article in the *Tampa Bay Times* featured an unnamed spokesman for the Industrial Engineering Corporation of West Palm Beach who said that 1,500 skilled workers would be needed to transform cabbage palms into wood and rope. The president of the operation was F. E. Getchell, and the article claimed both that there were 3.5 million acres of cabbage-bearing land in Florida and that the new facility in Dunedin, Florida, would produce restricted materials for the government:

"A new wood, reportedly made from cabbage palm trees, will be used to re-inforce existing Army cantonments and in the construction of new ones."[21]

Then, in 1953, the *Tampa Morning Tribune* reported that the Industrial Engineering Corporation of Tampa would be opening a palm products plant in Okeechobee, and, according to F. E. Getchell, the company wasn't asking for any concessions to locate there, "only the cooperation of the Chamber of Commerce in signing leases on land bearing cabbage palms in order to be assured a supply of raw material." The company planned to pro-cess 2,400 to 3,000 trees a day and therefore would need at least 400,000 acres under lease. While the company could make "gunpowder, paper, film, nylon, gum, broomstraw and other products," they were going to "concen-trate on the production of alpha cellulose, a product from which smoke-less [gun] powder is made." They had already made arrangements to can swamp cabbage.[22]

On January 25, 1957, the *Miami News* ran this headline: "3rd Suspect in Stock Deal Hunted in St. Petersburg." The two suspects in custody were F. E. Getchell and his son: "The trio is accused of organizing a firm, Florida Palm, Inc. which made representations of having a secret formula for pro-ducing quality paper from sabal and cabbage palm derivatives."[23] On Feb-ruary 9, the *Tampa Tribune:* "Three Deny Swindle in Okeechobee Palm Firm."[24] On March 4, 1958, the *Miami News* reported the trial was in its third week and that a patent office official had testified that three successive patent applications for a secret process to convert cabbage palms to paper had been denied.[25] March 15: Edward Fox of Daytona Beach testified that Getchell "told him ice cream could be made under a formula Getchell and his son Harry developed to capitalize on cabbage palm trees. Fox said he was also told the Shah of Iran was interested."[26] March 19: The head of the University of Florida's Wood Pulp Laboratory testified that cabbage palm pulp was slightly more than two-thirds alpha cellulose, not 88 to 96.2 per-cent, as Getchell's literature had claimed.[27] Investors and the government claimed the paper from the palm formula was worthless and investors had been defrauded. June 31: F. E. Getchell and three associates were sentenced to prison. Getchell was ordered to serve five, five-year concurrent terms. All four vowed to appeal.[28]

September 1960: The Fifth Circuit Court of Appeals cleared F. E. Getchell and the other three defendants. The court said the trial record "clearly shows the prosecution to be infected with prejudice, springing mostly from the extreme disappointment of disgruntled losing investors."[29] Did Getchell and the others deliberately defraud investors, or did they just have cabbage palm dreams that exceeded reality? Hard to say, but it's probably just as well. If F. E. Getchell had been right about the regenerative capacity of cabbage palms and the operation lasted sixty years, from 1958 to 2020, it would have consumed more than 54 million cabbage palms, using his low estimate of 2,400 per day.

With the exception of the contemporary landscape industry, the biggest commercial consumer of cabbage palms has been the brush-making industry, which managed to hang on into the 1950s, when nylon bristles eroded the market for natural bristle brushes. The two best-known brush companies produced Ox and Donax brushes. Ox Brushes were made by the Ox Fibre Company, which started in Sanford, Florida, and moved to Jacksonville, where it was renamed the American Fibre Company.[30] In 1901 some drying fiber caught fire, and the plant burned down, taking with it 130 blocks of Jacksonville.[31] The company then relocated to Benson Junction, next to DeBary.[32] The other big player was the Standard Manufacturing Company of Cedar Key, although there were clearly contributing operations, including one in Gulf Hammock: the "Old Fiber Factory" on Cow Creek, which is now part of Waccasassa Bay Preserve State Park.[33]

Daniel Andrews was an Indiana dentist who first visited the Cedar Keys in 1904 and started a palmetto fiber and brush-making business by collaborating with local resident and mechanical genius St. Clair Whitman. Together they imagined, designed, and built a complicated industrial sequence that converted young cabbage palms into fibers with the right balance of stiffness and flexibility to make excellent brushes.

The Standard Manufacturing Company began operating in January 1910. The enterprise was based on the ability of Andrews's men to go harvest small cabbage palms from coastal forests they didn't own, although they did pay a ¾ cent per pound "stumpage fee." The raw palmetto fiber was harvested from a region stretching from the mouth of the Suwannee River to Crystal River. Though only thirty miles by air from Cedar Key, collecting the buds involved carefully planned visits via hundreds of miles of tortuous tidal creeks navigable only by small boats at high tide. They called the harvested material "buds," but they actually consisted of nearly all of the aboveground portion of short cabbage palms, 5 to 8 feet long and weighing 75 to 125 pounds each.[34]

The crews used custom axes sold by the Collins Axe Company in Collinsville, Connecticut. These axes featured broad, thin, curved blades,[35] and today's serious swamp cabbage harvesters use a similar design. The most skilled "bud cutters" could harvest one hundred buds a day.[36] Oxen were used to haul wagonloads of buds to the banks of tidal creeks, where they would be stockpiled awaiting pickup by boat.[37] It could take as much as a month to accumulate enough buds to justify a boat trip to collect them.[38]

I've always wondered about the expression "need to light a fire under someone" as a motivational metaphor, which seems related to a story of a stubborn ox that refused to haul a load of buds. The frustrated driver actually lit a fire under the ox, but the ox only moved forward far enough to escape the flames, thereby parking the wagon directly over the conflagration and thus setting the wagon on fire.[39]

Once back in Cedar Key, the buds were cooked for three days in cypress tanks, and then the softened leaf bases or boots would be fed into the "head hackle machine," which allowed the bud to be separated into individual "boots." This must have been a monstrous machine that attacked a 100-pound palm trunk (that was euphemistically called a "bud") and disassembled it into its component leaf bases.[40] Then the softened boots (the individual leaf bases) were fed through a "boot roller" that functioned like an old-fashioned washing machine wringer to flatten the boots.[41] The lower, older leaves contained the coarsest fibers and the youngest, the finest. Next, each flattened leaf base would be fed into a wet hackle, a rotating drum with metal teeth that raked or combed through the cooked leaf. A stream of water helped wash away the nonfibrous material. The operator would hold one half while the wet hackle did its work, then pull it out, reverse it, and subject the half they had been holding to the toothy machine.

From the wet hackle, the bundles of fiber were taken outside and placed on drying racks, elevated chicken wire platforms, and then covered with more wire to keep the fibers from blowing away. From the drying racks the fibers went to an oil treatment house, where they were treated with paraffin oil, which appears to be what we now call mineral oil, that was believed to preserve the fibers and keep them flexible.[42] Then the finishing department dry-hackled them once more to remove any tangles, and the fibers were "drafted," which enabled workers to group fibers by similar lengths anywhere from 3¼ inches to 20 inches. In 1918 this palmetto fiber sold for twenty to twenty-five cents per pound, depending on the grade.

The vast majority of the palmetto fiber was sold to other manufacturers around the country, with only about 10 percent remaining to be made into Donax brand brushes in Cedar Key.

The Donax brand included a variety of brushes: whisk brooms (primarily used to dust off clothes), hat brushes, crumb brushes, an auto brush, a hearth broom, and a metal-handled utility brush. Dozens of outlets from Marshall Fields in Chicago to the Tampa Alligator Farm carried Donax brushes, and they were displayed at several world's fairs.[43]

According to the 1935–36 *Industrial Directory of Florida,* the Standard Manufacturing plant claimed to be capable of producing 1,500 pounds of fiber a day. Dr. Andrews's son cited a figure of two to three thousand buds a month (so figure one hundred buds a day), and at three pounds of fiber per bud, that would only be three hundred pounds a day.[44] That same year the Wooten Fiber Co. in Jacksonville claimed to be producing a million pounds a year, nearly nine times as much as Standard Manufacturing, but Wooten (also spelled "Wootten") was not making brushes but providing fiber for plastering.[45]

At that rate it seemed as though demand for palm fiber could eliminate Florida's cabbage palms. No one knew if any of Florida's palm-based moneymaking schemes would catch on, so there was real concern that manufacturing might threaten the existence of cabbage palms. Charles Sprague Sargent, director of Harvard's Arnold Arboretum, was quoted in 1924 by Henry Nehrling as follows: "One concern in Jacksonville, Fla., alone consumes 7,500 buds a week, [and] the time is not very far distant when the Sabal palmetto will become a rare tree. The buds, which are now mostly procured from the West Coast of Florida, where the palmetto is most abundant, are worth at the mill only 6 or 7 cents each, and it is not probable that the extravagant and wastefulness of this brush industry is exceeded by that of any other in the United States."[46]

In 1922 Daniel Andrews was issued two patents, 61357[47] and 61,358, for ornamental designs of brushes.[48] The Patent Office has been kept busy by a perpetual stream of people imagining new uses for cabbage palms. Several examples should suffice to demonstrate the range of obsessive ingenuity and vision deployed in the service of trying to convert acres of presumably useless cabbage palms into useful U.S. dollars.

On June 10, 1910, Frederick Strong of St. Petersburg filed a patent for a process of converting fibrous plants into textile fiber and pulp: "While my invention is not limited to the utilization of the fibrous plants of the tropics, but can be applied to various other approximately similar plants and growths, it is particularly intended for the production of crude paper pulp, textile and cordage fiber, etc., from the leaves, leafstalks, stemsheaths, trunk-wood, and roots of such practically inexhaustible supply as the cabbage palm or palmetto tree, the saw palmetto or scrub palmetto. . . . I have succeeded in developing the hereinafter described mechanical and chemical process whereby the somewhat coarse natural fibers of the plants are easily broken up into fine white fibers suitable even for spinning into fabric having the wearing qualities of linen and the luster of silk, and the cellular tissue is converted into a valuable pulp-product from which tough paper may be produced."[49]

In 1915, Frank Dexter of Vista, Florida, applied for a patent to make tires that avoided the use of rubber, and "A further object is to provide a tire formed primarily of endogenous growths such as the palmetto or cabbage palm (Sabal palmetto), the said material being arranged in segments with the fibers disposed along substantially radial lines whereby the ends of the fibers are caused to engage the surface on which the tire is mounted, thus to prevent slipping or skidding and at the same time producing a cushioning effect somewhat similar to that resulting from the use of a properly inflated pneumatic tire."[50]

On May 3, 1955, Eugene Brasol of Daytona Beach filed a patent for a cabbage palm flowerpot. "A flower pot adapted to contain soil and a growing plant, said flower pot comprising a shaped section of the trunk of the palm tree, *Sabal palmetto,* having the characteristics of light weight, durability, resistance to rot, decomposition and disintegration in the presence of moisture and being sufficiently porous to permit impregnation with a plant food and to retain for long periods of time the moisture in the soil which the pot contains."[51]

Cabbage palm–based patent filings didn't end in the 1950s. A 2006 patent mentioned *Sabal palmetto* as a possible "palm fiber-based dietary supplement."[52] And in 2016 Barbara Thorne filed a patent for termite control using materials from a Sabal palm tree: "In embodiments of the invention, materials from the *Sabal* palm tree may include portions, isolates, derivatives, extracts, suitable compost stages, or others of the *Sabal* palm tree. In embodiments of the invention, any part of the *Sabal* palm, either dead or alive, may be utilized."[53]

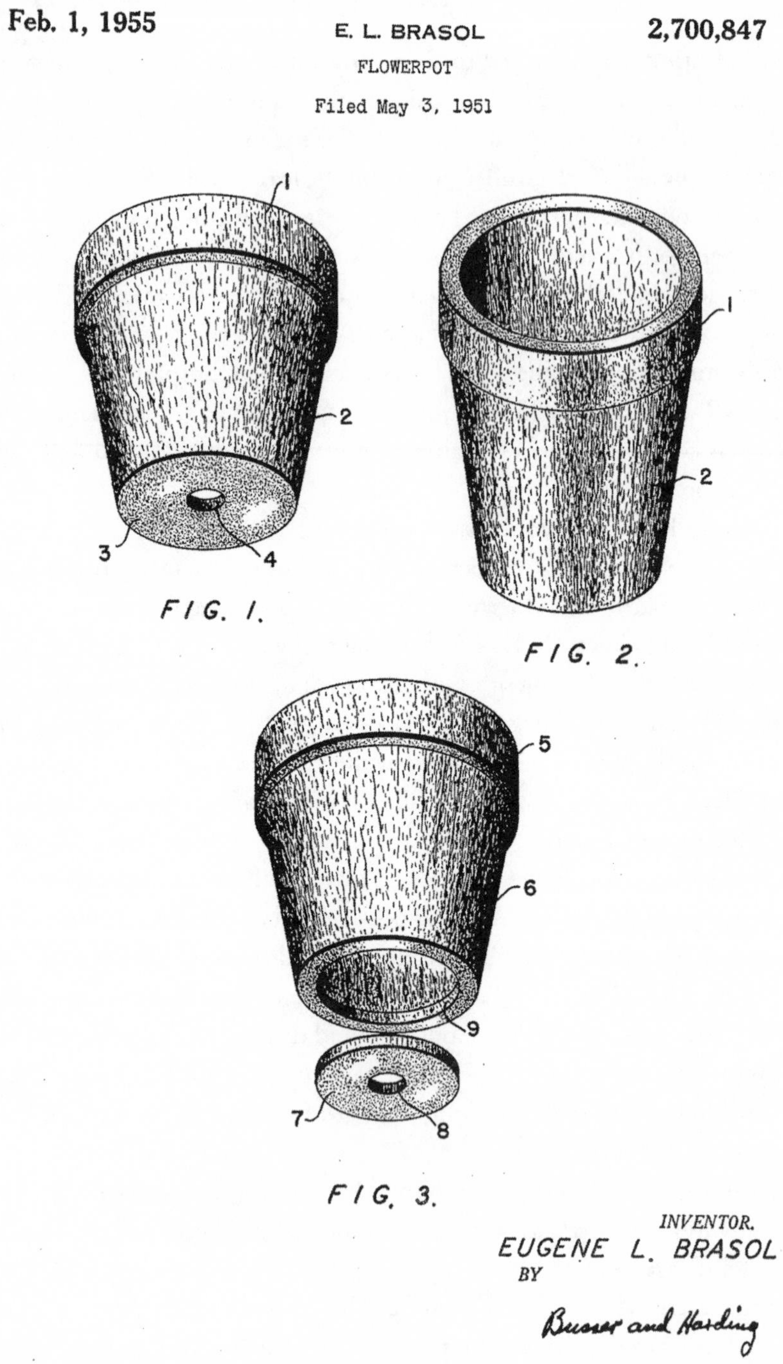
Feb. 1, 1955
E. L. BRASOL
2,700,847
FLOWERPOT
Filed May 3, 1951
FIG. 1.
FIG. 2.
FIG. 3.
INVENTOR.
EUGENE L. BRASOL
BY
Busser and Harding

Eugene Brasol's intriguing patent for biodegradable flowerpots made from palm logs. Source: United States Patent and Trademark Office, www.uspto.gov.

At times it seems as though the vision of what cabbage palms might actually do strayed fairly far from pragmatic reality. In 1948, the *Tampa Tribune* profiled Florida State Forester C. H. Coulter, by whose calculations there were three and a half million acres of cabbage palms growing in commercial quantities in the state, but only about a dozen firms producing cabbage palm products. Coulter wanted to change that and proffered a somewhat fantastical list of potential uses that affirmed he was a true believer—hot pads made from sections of logs, hats, mats, baskets, pin cushions, walking sticks, umbrella handles—and he noted that an oil cake for fattening cattle could be made from the seeds.[54] Other sources suggest cabbage palms might be used as pipes for carrying water, "ornamental table tops from polished stem cross sections," and canes.[55] In 1936 the *Palm Beach Post* suggested waterproof coffins might be made from palmetto fiber.[56] I suppose someone once made a tabletop, or a pincushion, or an umbrella handle from a cabbage palm, but I suspect these enumerated uses are perpetuated more as inbred references to prior citations than as frequently observed historic or contemporary uses. Currently, there is some sustainable frond harvesting for chickees, tiki bars, and religious purposes, and beekeepers take advantage of the nectar flow, but despite entrepreneurial zest, patent filings, and incurable American creativity, the only contemporary measurable economic consumptive uses of cabbage palms are in the landscape industry and some minor swamp cabbage harvesting. The number of cabbage palms lost to residential- and agriculture-related land clearing no doubt exceeds the volume of all trees otherwise consumed.

18

———•———

Seminole Thatch

Chickees and Tikis

I had driven from 5:00 A.M. darkness into daylight while avoiding raccoons and roadside deer to arrive at the Brighton Seminole Casino, a prefabricated building far larger than it was alluring. A retired and squinting couple emerged past a motor home while I laced up boots and slathered on sunscreen in the parking lot.

The first vehicle to arrive was a big dark pickup pulling a long trailer. I approached, and the driver's window slid down to reveal a handsome, weathered man who apparently spoke little or no English. As I looked up he gestured to the seat behind where another passenger informed me I should follow them to the site where we would be harvesting cabbage palm fronds.

We didn't drive far through the reservation and soon left the paved road to head north. We stopped at a cattle gate, and one of the four crew members hopped out to open it, and he waited to close it while both our vehicles passed through. After clearing the gate, I stopped, quickly tossed some of my front-seat gear in back and opened the door for a stocky, smiling man. I extended my hand, and said "Jono" as he got in. "Sandy," he said.

"Sandy?" Koufax notwithstanding, most of the Sandys I've encountered have been Anglo women, and nothing about this guy read as a Sandy. "Yes, you know, Sandy, Domingo." Ah, Sunday: Domingo.

Domingo and I followed the trailer through a small patch of hammock and emerged in a grassy prairie/pasture. Now, my two-wheel-drive Ford Ranger is great for yard sales, schlepping canoes, and helping people move boxes across town. But when bumping along behind someone off-road on a ranch, it engenders a particular anxiety. If everything goes well, no one will notice your inadequacies. But if not, the big boys will have to bail you out. The problem at Brighton isn't deep mud or dry, tire-spinning sugar sand, but ditches, and the landscape we were driving through had a reliable north-south ditch incised every 200 feet. A hundred and fifty years ago the palm hammocks we were heading to were probably an archipelago in a flat, wet prairie—islands in an intermittent freshwater sea flowing to Lake Okeechobee. The ditches weren't deep—ambitious linear scrapes, really—but they did their job and dried the landscape to allow for cattle grazing and more upland diversity in the hammocks.

We parked next to a small hammock island as two sandhill cranes stalked off through the pasture. The oval hammock's axis ran from the northwest to the southeast, and I learned later from Google Earth that the hammock was about 460 by 320 feet. While the other men unloaded three aluminum extension ladders from the trailer, the more conversant member of the crew explained that since there were about 200 fading fronds in the trailer, today's work would consist of gathering 2,800 more. Since at home I cut green fronds off my palms with a battery-powered reciprocating saw, I started trying to calculate how many weeks it would take me to accumulate 2,800 green palm fronds. It was nearly eight in the morning, and I was not optimistic.

Without much discussion the men grabbed an orange plastic Husqvarna chainsaw scabbard, and each withdrew a machete. One cleaned his by rubbing it vigorously across a sandy fire ant mound.

It seemed obvious that I should tag along with the most outgoing member, who, armed with just a machete, shouldered a ladder and plunged into the hammock while attempting to explain the basics. He was personally responsible for accumulating 700 fronds. They could only use big green fronds that hadn't been chewed on by cattle. Extending the ladder with a chorus of ratcheting clangs, he told me his name was José and that he had been gathering fronds at least once a week for eight years. Turning to an iPhone app, I informed him that would be a minimum of 291,200 fronds, assuming 700 was the usual count.

He placed the ladder quickly, but not casually. In addition to checking the trunk for any perceived weakness, he also looked aloft as the ladder made contact. If the ladder's sudden presence in the canopy revealed any startled wasps, he chose another tree. He was clearly proud of his arboreal skills, and he claimed the higher leaves are bigger and in better condition. He said he had never fallen, and that boast initially seemed unlikely. José has probably placed a ladder against something like fifteen thousand skinny palm trees, and one would think any imbalance at all would quickly lead to the ladder rotating around the cylindrical trunk, resulting in an unavoidable fall. But José doesn't place the ladder against the trunk—he impales the top rung or rungs on protruding leaf bases, and these are the same leaves that remain anchored in hurricane winds.

He climbs quickly and dispatches any dead fronds blocking his way. Then he sets to work with the machete, and most fronds fall with a single whack. Some take two, but it is surprising how fast 15 to 20 usable green fronds fall to the ground. He cuts close to the tree where the fronds are stable, so many of the falling leaves are more than 6 feet long at this point. Once the tree is shorn, José courteously checks my whereabouts, tosses the machete to the ground, and descends. Only three or four green leaves remain on the palm when he's done, and José asserts the canopy will grow back in three or four months. I'm dubious, but he has cut this hammock before, and I bite my tongue since I'm trying to learn about his understanding of these palms, not download mine.

Aloft again I note that he doesn't hold the leaf he is hitting with the machete. Instead he hangs on to the palm and not the ladder. But if that slashing hit counts as a touch, it is just a first in a long series of touches that a frond experiences before it is fixed in place on the roof of a chickee. Once the ladder is down, José collects the leaves and stacks them loosely but neatly, right side up, gathering the leaf stems (petioles) together not unlike a giant bouquet. Each frond also receives a decisive stomp in an attempt to convert the dramatically three-dimensional frond into a flatter aspect. After he has worked a few trees, he switches back to these stacks of cut fronds. He shoulders what seems like a couple dozen at a time and walks them out of the hammock to the edge of the prairie. I suspect my questions are slowing José down, so my pitching in is an attempt to offset our chatter. I'm able to handle about twelve fronds. In addition to their being slippery, I quickly find that standing on the tips of the fronds complicates the process of lifting. If you've ever stood on a shoelace and attempted to move the other foot, you know the effect.

Taking two fronds at a time and holding the petioles close together,

José uses one whack to shorten the long petioles to a working length that approximates the length of his machete. José explains that the horizontal supports that will ultimately hold the fronds are 18 inches apart, so this second cut reduces the fronds to a more appropriate length. Once they're trimmed, José creates a new pile, and now he is counting. After roughly two hours his tally is 255 fronds, which is somewhere around two a minute—an impressive rate of production considering all the steps involved.

I wander off to explore the hammock. Most of the palms are quite tall, well out of reach of any 24-foot extension ladder, but I'm surprised to see how many trees have been uprooted. Upon inspection I find the bottom of each root mass is not ragged, but flattened—imagine the underside of a mass of spaghetti congealed on a glass plate. Spying a shard of smooth rock, I realize these trees are rooted in just a few feet of soil above limestone. Their roots can spread outward for many feet but not downward. Some storm (Hurricane Charley?) ripped through here and tipped a number of trees over. Guavas and sour oranges are non-native elements, spread by either wildlife or Seminoles, and probably both. Stoppers, beautyberry, and hackberry are common, but you'd have to call it a cabbage palm/live oak hammock since those two species dominate the canopy.

Evidence of cattle is everywhere in the hammock, and they appear to be having some digestive problems so I'm watching where I step. Something the size of a tennis ball can lid catches my eye on a palm log. It is a coiled pygmy rattlesnake, something José warned me about. The experts say their bite is seldom fatal, but that's the good news. I took some pictures and left it alone. When I showed José the picture he suggested we go back and kill it because of the risk to the cattle, and before we left, the crew had killed two others. I went back several times to check on my small specimen, and it was in the same location each time. I rationalized that sparing this tiny fellow was helping me accrue some karmic capital that will protect me as I reach down to gather fronds from the ground.

In addition to rattlesnakes and wasps, José has to contend with fire ants, which are common in the prairie pasture but rare in the hammock. Falling is always a risk, but the biggest threat is making a mistake with the machete. He tells me one of the men nearly cut his hand off and had to be evacuated but now is working again. When I tell him that palm trimmers in suburbia use chainsaws, he looks at me in disbelief. He can't imagine using a heavy chainsaw instead of a machete. Apart from the whacking, the hammock is very quiet, dampening the sounds of the cows and meadowlarks in the pasture. I shudder imagining what four chainsaws would sound like.

Domingo, Joaquin, and Sebastian, the driver whom they call "the Chief," have focused on another nearby hammock about 500 feet away, and their technique is evidently less ladder-dependent since they have but two ladders for the three of them. They focus more on palms with canopies within reach of the ground. José explains that they live in Immokalee, while he lives in LaBelle. All four are Mexican, and Sebastian is called the chief because he is Mexican Indian. On more than one occasion Sebastian has been approached by chatty Seminoles or Miccosukees who have assumed he is of Creek descent. The Mexicans find this amusing. Without my asking, José volunteers that he doesn't think many Seminoles are building chickees or tiki-huts anymore. We agree casino money is the likely explanation and drop the subject. Chickee book author Carrie Dilley tactfully explains: "As tribal members have seen an increase in financial stability in recent years due to gaming revenue, it has become more difficult for chickee-builders to find willing laborers within the tribe. This has caused builders to look elsewhere for help, often leading them to hire non-Indians for the harvesting and constructing processes. Even when non-Seminoles work on the construction crews, the structure created will still be considered an authentic chickee as long as the owner of the chickee building company is Seminole."[1] Other legislation prevents non-Seminoles or non-Miccosukees from using the term "chickee," which contributes to the proliferation of tiki-huts.[2]

At quarter to twelve, José is up to 550 fronds, and the others arrive to speed things along. Evidently their ground game moves faster than José's ladder work, or maybe I've been slowing him down more than I thought. Although my contribution was minimal, having all five of us working quickly racks up the remaining 150, and we break for lunch. José and I sit on palm frond sit-upons, while the Immokalee men converse in Spanish nearby. I learn that the trees landscapers see as sabals and crackers see as swamp cabbage are simply "palmas" to the crew.

Having reached their quotas, the endeavor enters a new phase. The trailer is to be driven around the hammocks to pick up the piles. Joaquin and Domingo hand palms in bunches of five or six to José, who lays them carefully in the trailer and also continues to stomp them. Despite his efforts, the load is buoyant, and I can finally see how sleeping on boughs

Fresh green fronds destined for chickee thatch are loaded for travel. Photo by the author.

might approximate comfort. Sebastian does the driving and joins me in the pickup bed that is brim-filled with thousands of trimmed palm petioles and cypress scraps—leftovers from the last job. Silently we fling the biodegradable waste into the hammock. When he heaves a heavy plastic bag off the truck, I wince inwardly, but before driving on, he descends, rips the bag open, dumps a load of sawdust and places any water bottles and soft drink cans and the bag back in the truck.

The trailer apparently held about 3,000 cabbage palm fronds. The level of fronds extended up over the sides when they were done loading, but the weight of the ladders and some serious strapping compressed the load into a form that survived the drive to Sarasota.

Contemporary chickees are comprised of cypress poles and palmetto fronds and a lot of nails. Before nails were available, the thatching was tied in place.[3] The joinery is not meticulously fitted, but substantial—this is rustic construction. As José described, the horizontal purlins that the fronds are nailed to are 18 inches apart. Each frond is folded and attached to two purlins, one nail in each, the petiole or leaf stem pointing up slope. They are spaced horizontally about 3 inches apart. The most meticulous builders space them three fingers (2 inches) apart, which increases both the cost and the number of fronds needed, but which lengthens the life of the roof, which can vary from five to fifteen years.[4] In some chickees all the fronds are folded in the same direction; others alternate the fold for each course. Helpers on the ground use a stick with a nail protruding from the end to spear leaves and supply the men on the roof. As with any roofing project that relies on overlapping layers (such as shingles), the work starts at the eaves and proceeds to the ridge.

Plate XXXI[5] of the Theodor de Bry engravings based on LeMoyne watercolors,[6] which predate the Seminoles by a century and a half, clearly show palm-thatched roofs in what is now Florida. Historically, cabbage palms played an even larger part in chickee construction than they do today. Clay MacCauley's 1887 *The Seminole Indians of Florida* describes a typical chickee shelter as "only a covered platform," being approximately 16 by 9 feet "made almost altogether, if not wholly, of materials taken from the palmetto tree." "Eight upright palmetto logs, unsplit and undressed, support the roof."

The chickee covered a platform about 3 feet off the ground. "The platform is peculiar, in that it fills the interior of the building like a floor and serves to furnish the family with a dry sitting or lying down place when, as often happens, the whole region is under water." The platform was made from palmetto logs split transversely with the flat sides up and "lashed to the uprights by palmetto ropes, thongs, or trader's ropes."

"The thatching of the roof is quite a work of art: inside, the regularity and compactness of the laying of the leaves display much skill and taste on the part of the builder . . . the mass of leaves of which the roof is composed is held in place and made firm by heavy logs, which, bound together in pairs, are laid upon it astride the ridge."[7] Evidently making the ridge waterproof is somewhat of a challenge, and today some chickees are not afraid to sport a layer of tarpaper atop the ridge, frequently held in place by the aforementioned crisscrossed logs.

The floor of a traditional chickee was dirt, which typically means sand. The space inside the chickee is dense shade, which during daytime usually contrasts with the bright sunlight outside. As a result, illumination inside the chickee is contrary to conventional structures since sunlight reflecting off the sand illuminates the undersides of the timbers, casting shadows upward rather than downward.

Chickee thatching proceeds from the eaves toward the ridge. Photo by the author.

A hundred and twenty-five years later the chickee has evolved from an essential flood-avoiding shade structure in the Everglades to an urban or suburban indulgence featured in exotic-themed restaurants and backyard "tiki huts." Due to Florida's fondness for appropriating anything tropical, thatched structures built by Floridians inevitably seem to end up being portrayed as Polynesian, rather than Native American. Actually, the whole

American tiki culture is imported from California, where it was developed at themed restaurants. The appeal of the tiki hut/chickees in Florida seems to lie both in their rustic, if bastardized, charm as well as the suspicion that building one is simultaneously connecting one with native Floridians while flouting some onerous laws. That's because commercial chickees constructed by Seminole or Miccosukee operators are exempt from "the Florida Building Code 553.73, Section 8 part i: The following buildings, structures, and facilities are exempt from the Florida Building Code as provided by law, and any further exemptions shall be as determined by the Legislature and provided by law: Chickees constructed by the Miccosukee Tribe of Indians of Florida or the Seminole Tribe of Florida. As used in this paragraph, the term 'chickee' means an open-sided wooden hut that has a thatched roof of palm or palmetto or other traditional materials, and that does not incorporate any electrical, plumbing, or other nonwood features."[8] While they don't have to comply with the building code, chickees do need to comply with zoning codes.[9]

The term "tiki" can apply to five different things in Florida: (1) the Anglo interpretation of a chickee—an open-sided, typically liquor-dispensing, palm-thatched hut, (2) Disney World's Enchanted Tiki Room, laden with chatty, musically inclined audioanitromic birds, (3) the identical twin brother of former Tampa Bay Bucs cornerback Ronde Barber, (4) the backyard torches (with the liquid fuel unfortunately packaged like apple juice that can easily poison small children),[10] and (5) the vertical log carvings. As with the tiki hut, the tiki carvings are also Polynesian by way of California. The first Floridians did make totemic wood carvings, but this tradition was apparently not adopted by the Seminoles. Nevertheless, tiki huts are frequently accessorized not only with copious amounts of bamboo, floral imagery, and rattan furniture but also with stylized tiki carvings.

Chainsaw carvers are found throughout Florida. Some work with cedar, some work with pine, and some carve standing dead oaks in place, but the preferred wood for tiki carving appears to be palm, and cabbage palm predominates. Some sell from the outdoor studios where they work, some have websites, and some simply set up and sell alongside the road.

Inspired by a Florida honeymoon in the late 1970s, Tiki Tim moved to Florida in 1980 burdened with drug and alcohol addictions, a failing marriage, and a propensity for earning a living dealing drugs. Despite being born into a poor family with two deaf parents living on welfare while his father spent time in Elgin State Hospital dealing with schizophrenia, Tim characterizes himself as having had a normal childhood in Chicago. When I chatted with him he was going on five years clean, single, and had become a self-employed entrepreneurial artist who works only five hours a day and routinely turns away offers for more work.

His upward path has been forged from a lifelong interest in art he re-engaged while serving a year and a half in prison. After doing his time he worked in marine construction until he tore his rotator cuff. That led to legal cases regarding his medical expenses and being fired. He spent nine months recovering and in therapy, during which time he started making small carvings out of Florida red cedar. People expressed interest in these small carvings, which he enjoyed making, but he couldn't see how to translate tiny figures into something that might support him. When he was fit to work again, "Nobody wanted to hire me . . . seven-time felon with several workman comp cases."

"So out of desperation I picked up a chainsaw, which I learned to handle building boat docks." He had always been interested in Polynesian art, and he decided to make himself a tiki. He started by carving a half dozen tikis out of cut-off sections of dock pilings. The first day by the side of the road he made $350. That roadside retail success opened a door. This first foray into tiki carving coincided with getting in a program to deal with his drug and alcohol problems, which was both fortuitous and essential: "I mean without that . . . and I run a chainsaw all day long . . . see me running around with missing body parts if I didn't have some kind of program."

The cylindrical green dock piling leftovers were familiar material, but the toxic chromated copper arsenic used to preserve the pilings was giving him nosebleeds. A veteran chainsaw tiki carver (Rod Green) told him he had to stop using that stuff, and so, knowing traditional tikis were carved from palm logs, Timothy gradually identified local sources of palm logs, which are the by-products of local land-clearing operations.

"It was like trial and experiment . . . on what kind of palm trees to use. And I found out that cabbage palm was completely different from most palms. . . . I tried queen palm, I tried royal palm . . . looks like Styrofoam on the inside, you would think it is a dense wood. . . . Some of the date palms are carvable. . . . Of course, coconut tree, but it has to be the very thick. . . . The sabal palm, I shouldn't say it's consistent. It's kinda like

Forrest Gump, you know, you never know what you get till you open one up, ya know."

Tim now identifies himself as a Floridian and is living the Florida dream. His livelihood depends on a steady supply of local native palms displaced by land-clearing operations. On weekdays Tim can be found in his backyard using a special crib to hold palm logs horizontally for his chainsaw sculpting. On weekends he heads out to roadside locations where he parks his truck and sets up his display. He has been run out of North Port, Sarasota, and Bradenton, so when I talked to him he had been alternating between a spot on Southbound 19 along the Skyway Causeway and another causeway location on the west side of the Gandy Bridge.

Virtually none of the other products sold from the roadside or abandoned filling stations in Florida are the work of the vendors themselves. When I explored this possibility and opined, "Tiki carvers seem like the only people who make their own stuff," Tim promptly responded, "They were the original Highwaymen." Tiki Tim had just made an unsolicited connection to the Highwaymen, Florida's black artist entrepreneurs who sold work they created, either door to door or from the trunks of their cars. Like the paintings of the prolific Highwaymen, Tim's tikis simultaneously meld organized production techniques with individual uniqueness. "I carve 'em . . . you know . . . after you have been doing it for a while, you know, it's like a repetition, the more you do it the faster and the more . . . the better you get at it. It's just like anything."

19

———•———

Sweetgrass Baskets

Gullah/Geechee South Carolinians Create Value from African Skills

Charleston and two neighboring communities north of the confluence of the Cooper and Wando Rivers, Mount Pleasant and Sullivan's Island, are the epicenter of palmetto obsession in South Carolina. This is where palmettos grow well, where they first made Carolinian history, and where palmetto leaves are crafted into two iconic products that are uniquely associated with the region: palmetto roses and sweetgrass baskets.

At the City Market in Charleston and elsewhere one will encounter the purveyors of the palmetto rose, which is a passable simulacrum of a rose blossom rendered in dried palmetto leaflets. At three dollars or less, it is an affordable gift that, being dried, lasts for years and therefore serves as an enduring souvenir of a visit to Charleston. With two leaflets in hand, a skilled practitioner can create a palmetto rose in about a minute. Once one is shown how, they are easy to make, and in some communities adjacent to a ready supply of short, within-reach palmettos, the homeless offer them for free on sidewalks, although a "donation" is an ethically implicit quid pro quo. In addition to roses, crosses, fish, and grasshoppers are sometimes seen, but this selection is greatly diminished from the nineteenth century, when hats, baskets, brushes, and fans were made from palmetto for personal use and the tourist trade.

As Charleston is the home of the palmetto rose, across the river Mount Pleasant has become the official home of the sweetgrass basket. Although the Sweetgrass Basket Pavilion[1] is more or less under the dominating

Ravenel Bridge, the sale of these baskets in Mount Pleasant evolved along Route 17, where vendors tried to catch the eye of motorists. This roadside-sales approach apparently followed the 1929 opening of the first span linking downtown Charleston to Mount Pleasant when Lottie "Winnee" Moultrie Swinton is said to have placed a chair next to U.S. Highway 17 in Mount Pleasant and started the practice.[2] Since 17 was the coastal highway of choice prior to the interstates, thousands of travelers had multiple opportunities to pull off and inspect the wares.

The custom of roadside sales from the right-of-way or vacant land not owned by the vendors appears to be a vanishing southern tradition. You can still find peaches, watermelons, and shrimp sold from the back of pickups, as well as fluffy alpaca rugs at gas stations. And of course there are the seasonal firework, pumpkin, and Christmas tree lots. But unlike sweetgrass baskets, these products are typically bought from wholesalers and are not made by the vendors themselves.

When you buy a genuine sweetgrass basket from the Lowcountry of South Carolina, you are purchasing only two ingredients: sweetgrass, *Muhlenbergia filipes* (and sometimes pine needles), and palmetto, all wild-harvested, natural, and sustainable products. No wire, no glue, no staples, no shellac, no pins, no hidden armature, no threads, no stabilizers, no filler, and no nonsense. Just a coil of sweetgrass bundles spiraling around and around, constrained and connected by helical binder made of young palmetto leaf that stitches one coil to the next. So what we know as a sweetgrass basket is more accurately a sweetgrass/palmetto basket—for without the palm binders, there is no basket. Sweetgrass (once called seagrass) has a cachet in coastal South Carolina that sometimes eclipses even palmetto. One can buy a home in the Sweetgrass housing development, and I've even seen palmetto roses sold as "sweetgrass roses" in the City Market in Charleston, despite the fact that there is no sweetgrass in the product being offered.

Constructing a sweetgrass basket requires skill several orders of magnitude removed from what is needed to make a palmetto rose, and though, with time, virtually anyone could learn to make these baskets, virtually no one does. No one but members of the Gullah/Geechee community in the Lowcountry, where sweetgrass basket construction is a folk skill passed from one generation to the next. The technique came from Africa and was crucial for rice culture—something the slaves knew far more about than

Lowcountry sweetgrass basket detail. Photo by the author.

their masters. After harvest, the rice had to be threshed, the process that detached the grains from the hulls using a mortar and pestle.[3] The mortars were sometimes constructed from palmetto trunks.[4] Threshing did two things: it separated the hulls from the grain, and it also "polished" the brown rice into white rice by removing the rice bran[5], which is nutritious, but predisposes the rice to becoming rancid. But even after being threshed, one can't eat an admixture of rice grains and hulls, so lightweight "fanners" (large, flat baskets) were used to separate rice grains from the hulls by flipping the rice and hulls into the air in a manner that allowed the lighter hulls to be blown out and the grains to land back in the fan. African rice fans were constructed of different plant materials, so the slaves adapted their technology to what was available in coastal South Carolina. The original utilitarian fanners were frequently made from black rush "bound with thin splits of white oak or thin strips from the stem of the saw palmetto."[6] So today's sweetgrass basket makers inherit an evolved legacy of rice-based culture implemented by slaves, and the amount of patience and practice embodied in a good piece is reflected by the price, which can be considerable. A sweetgrass basket is therefore simultaneously a utilitarian object, a souvenir of the Lowcountry, a work of art, and an affirmation of the re-sourcefulness and skill of the Gullah/Geechee community whose ancestors brought with them from Africa basket-making skills that proved invalu-able in winnowing rice. This places roadside vendors of sweetgrass baskets in a completely different league from those hawking velvet paintings, and it explains why these baskets are in collections such as the Smithsonian.

So I was somewhat aghast when south of Charleston on Highway 17 in a shop that appeared to promise authenticity, I saw what looked like a sweetgrass wastebasket for twenty-five dollars. This was no textured plastic wastebasket, but a legitimately handmade construction of grass and palm. Upon closer inspection I realized the palm was the outer layer of the rat-tan palm, which is commonly called cane. There is no rattan palm grow-ing in South Carolina. If you have ever looked closely at a cane-bottom chair, you've noticed the distinctive crackly appearance of the cane. That's what I was staring at in the knock-off wastebasket. Superficially, it looked like a sweetgrass basket, but the materials came from and were assembled by someone on another continent with no connection to the Lowcoun-try, rice, sweetgrass, palmetto, or American slavery. I was holding a com-mercialized cultural appropriation, which was truly regrettable since many cultures on the other side of the planet excel at their own indigenous palm-craft.

Cheap imitations are just one of the threats facing the sweetgrass basket makers. In 2013, historic Highway 17 running north out of Mount Pleas-ant was "improved" by widening and adding curbs. This was ostensibly a safety-inspired change, but it prevents potential customers from simply pulling off the highway to patronize a stall. As a result, the stalls are now tucked aside in shopping center parking lots, and adjacent to housing de-velopment and church entrances.

For eight decades the craftspeople, mostly women, along Highway 17 went about their business largely unregulated and unrecognized by govern-ment. But now sprawl and the government road-widening have taken note of the basket makers—honoring them with special designations (Official State Craft! Sweetgrass Basket Makers Highway!), and even permitting considerations (Sweetgrass Basket Overlay District![7] and transportation impact fee credits!)[8] while relegating the artisans to the marginalized cor-ners of commercial highway sprawl. So instead of the legacy of Gullah/ Geechee culture dominating the neglected Highway 17 roadside as it had for years, it has been compressed and confined—what for many decades had been the roadside entrepreneurial expression of a culture is now increasingly anachronistic—nearly lost amid IHOP, Lowe's, Harris-Teeter, Ethan Allen, and P. F. Chang's.

The suburbanization of the Lowcountry hasn't been restricted to strip commercial along Highway 17. The privatization of the landscape brought about by waterfront real estate and gated communities has limited access to the natural resources needed to make the baskets. The most pressing need is access to sweetgrass, a species of *Muhlenbergia* that grows in a relatively narrow zone between salt marsh and the maritime forest. The sweetgrass is not mown or cut, but pulled, and it is believed this thinning helps keep the grass from choking itself out. Efforts are under way to cultivate the sweetgrass to assure a future, but basket makers and their suppliers can't help but lament the switch from nature providing to having to provide for nature. As if sweetgrass supplies weren't problematic enough, these baskets also depend on a supply of young palmetto leaves, and that's where people like James Moultrie fit in.

I encountered Mr. Moultrie south of Long Point Road on Highway 17, where he was balancing somewhere between fifty and one hundred palmetto leaves on the handlebars of his bicycle as he peddled south. He has been selling palm leaves to the basket-crafters along Highway 17 for two decades, and many of the basket makers I spoke with knew him. When I met him in 2013, the going price was two dollars per leaf.

Mr. Moultrie uses a small paring knife to harvest the unfolding leaf, not the central spear leaf. His approach is congruent with the advice offered in a 1947 book on palmetto braiding and weaving, which instructs: "The leaf next to the bud leaf of the palm should be cut as soon two or three inches shows above the wrapper or boot. (Cutting this second leaf from the tree does not injure the palm.) At this time in its development, the leaf is folded tight and pressed against the bud leaf, with no green except on the edges of the fronds." The authors go on to suggest: "The best practice, from the authors' own experience, is to gather at the dark of the moon."[9] No explanation was offered, but at a minimum this practice would seem to limit harvest to thirteen leaves a year, a number at odds with Mr. Moultrie's assertion that he can go back to the same tree every two weeks, which would suggest the palmetto has the capacity to produce twenty-six new leaves a year, a rate of production I found impressive, if unlikely.

James Moultrie supplies sweetgrass basket makers with palm leaflets. Photo by the author.

Traveling by bicycle, Mr. Moultrie is both pedaller and peddler. He sells to the basket makers along Highway 17, who then have to process the young palmetto leaves into usable binders. Used green, they will shrink, so some drying is necessary to stabilize the material, but they can't be allowed to dry too much either. The leaflets are separated and the tougher rib-like fibers (the butt) stripped off.[10] The resulting strips, shorn of their butts, are even thickness, and they are, perhaps counterintuitively, widest in the center, probably never more than ¾" and taper to either end. These are then split to create weavers anywhere from ¹⁄₁₆" to ¼" wide. These binders are typically stored in a large plastic bag. In order to be used, one end needs to be trimmed to a point so it can pass through a hole made in the sweetgrass coils. The holes are formed by "nail bones," which nowadays are frequently filed-down teaspoons that are thrust through the collected sweetgrass leaves, creating a space for the binder to link the preceding course to the one being created.

Part of the appeal in buying a sweetgrass basket is dealing directly with the artist, asking questions, watching her, or him, work, coming to understand and appreciate the process, and maybe snapping a picture or two. Some artists are willing to negotiate, but even with some give-and-take, these are not cheap souvenirs. Short on cash, I bought a hair barrette for half off, but the asking price was forty dollars. In 2015 a Charleston-based basket gallery website was selling nine baskets for more than $1,000 and one was $24,000. A "towel/show" basket 28 inches high and similar to the foreign knockoff I'd seen for twenty-five dollars was priced at $1,690.[11] Of course, with a twenty-first-century online gallery transaction the buyer doesn't get to chat up the basket maker, which is part of the experience.

These skilled artisans endeavor to pass their skills to the future generations with mixed success. In 1971, one basket maker opined: "This is just too much monotony for the kids today. They don't want to sit all day and half the night and weave with their fingers and a little sawed-off spoon for a shuttle because they can make more money doing almost anything else."[12] A 2012 article reported some basket makers earned "$17,000 a year or more,"[13] which could be a nice supplement, but would seem to argue against basket making as a sole profession. Yet since not all basket makers today are over forty-five, the tradition persists despite being overtaken by highway strip development, being separated from their vanishing historic sweetgrass gathering areas, and being threatened by foreign imitators.

South Carolina's contemporary sweetgrass/palmetto basket makers have transformed a relict of their ancestors' oppression into a high-value product made from a few natural materials, time, an old spoon, and more than three centuries of handed-down tradition.

20

---•---

Palm Abodes

A Different Approach to Log Cabins

Cabbage palm log structures have appeared throughout the Southeast, but it is hard to know when they first appeared. If you visualize Native American woodworking tools being limited to mollusk shells, stone, and fire, it is challenging to accept the premise that they were felling significant numbers of palm trunks, because cabbage palms are notoriously hard to cut down and they don't readily succumb to fire. On the other hand, eroding beaches and riverbanks produce a continuous stream of palm logs, and anyone willing to lug them around could have easily built rudimentary log cribs of horizontal logs or buried them in vertical orientations. Still, if North Americans were building shelters from palm logs before Spanish axes arrived, we have no definitive record of it.

It is easy to visualize how someone seeking shelter under a broad waterproof palm frond during a deluge could lead to a hypothesis that the leaves might be useful for thatching roofs. Chief Billie of the Seminole Tribe of Florida has said: "A cabbage palm frond is a fine umbrella, but a tree is better still if you find yourself in the rain. Just lean back against the trunk and watch the rain go by. You'll stay dry. A rain shower in the woods is a good time for dreaming under the arms of the cabbage palm."[1]

In 1539, a member of the De Soto expedition, having arrived at a village called Ucita, described "houses built of timber, and covered with palm-leaves."[2] The earliest extant European images of Florida are the late sixteenth-century LeMoyne/de Bry engravings of Florida that, unfortunately, were drawn from memory of a 1564 visit and depict crudely rendered palms and Native American encampments surrounded by "thick round palings the height of two men."[3] Whether these were pine, palm, or cypress logs is impossible to discern from the engravings. Plate 30 of the series shows how flaming arrows were used to ignite the thatched roofs of enemy tribes.[4] Despite the appeal of these early depictions, the details and even the existence of these palisaded encampments have been questioned.[5]

European and American settlers adopted palm thatch as an expedient, if flammable, approach to roofing, and the creation of palm-thatched buildings has persisted to the present. The best known and best crafted are the chickees built by the Seminole Tribe of Florida, the Miccosukees, and the Independent Seminoles. But early white settlers, outdoorsmen, and survivalists have all recognized the waterproof utility of palm fronds. The advent of photography made it clear that white settlers in Florida routinely used palm fronds for both shack roofing and exterior walls.

The arrival of metal axes and saws in the New World no doubt led to an explosion of log construction. At first, continuing the trend depicted by de Bry, fortifications may have been a priority, but it was inevitable that in addition to fort walls, other structures would eventually be fashioned from palm logs. So far I've identified more than two dozen vanished or existing examples: in Ormond Beach, Palmdale, along the Silver River, at St. Petersburg, and the Homosassa/Yankeetown area. I wouldn't be surprised if there are another dozen out there. Surviving photographs of cabbage palm residences show that some utilized vertical logs, others were placed horizontally, like traditional log cabins, and at least one utilized horizontal log slab sections stacked like cordwood. Consequently, I suspect palm log cabins were reinvented independently around Florida, but there are also instances of one cabin having inspired others. Perhaps the most surprising thing about cabbage palm log structures is their durability.

A century and a third ago, carpenter craftsmen began work on what appears to be a rustic Adirondack lodge, an eight-room, two-story edifice with a corner fireplace in each room. It was built as part of the 80-acre Santa Lucia Plantation by Portland, Maine, native John Anderson. Native materials and rustic features epitomize the Adirondack architectural style, and this curious Florida example follows suit—native materials included massive 11" × 11" heart-pine post timbers, small cedar branch railings and walls of intact palmetto logs. The one-log-thick walls were chinked with both Spanish moss and the russet fibers from the base of the palm fronds.

This two-story log home, Talahloka, is the oldest occupied residence in Volusia County. Postcard. Author's collection.

It was one of at least three palmetto log buildings built by Anderson. The home is essentially a two-story square with a hefty central chimney that accesses the innermost corners of the four rooms on each floor. The pyramidal hipped roof overhangs a second-story veranda. In the current configuration the four downstairs rooms are connected, and the owner uses a narrow interior stairway to access a bedroom and bath. The other three second-floor rooms are accessed by an exterior stairway on the west side. The National Register documentation opines, "The use of palmetto log construction, however, is unusual because of the soft fibrous consistency of the wood which makes it generally unsuitable for construction."[6]

So, paradoxically perhaps, this rustic dwelling, ostensibly (according to the National Register) made of the least substantial and generally unsuitable material, is the oldest occupied structure in Volusia County due to three factors: the sheltering roof that kept elements 6 feet from the palm log walls, the general durability of old palmetto logs protected from the elements, and the dedication of the current owner, Sang Roberson.

No doubt it would have been pushed over had Sang and her former husband not purchased it in 1998. Over the years the former hunting lodge (and rumored bordello) has contended with squirrels in the eaves, woodpeckers and screech owls nesting in the log walls (owners covered the holes with sheets of tin), powderpost beetles, and termites. The termites have mostly affected conventional lumber, but I convinced Sang that dry-wood termites do sometimes go after cabbage palms.

Sang is an accomplished potter with a summer home in Taos, New Mexico. Our first meeting had been postponed several times due to her full schedule of studio clay work, grandchildren, guests, and historic preservation activism. She is neither young nor tall but embodies a southern grace and competence that allows her to unself-consciously refer to others as "little old ladies." Sang is the structure's protector-in-residence—she's had it tented twice for termites and made a number of modifications, some remedial, some requisite functional upgrades. She's contended with sagging subfloors and settling brick piers; the porches have been rebuilt, and all the rustic twig-style railings replaced with modern lumber. What hasn't needed replacing or buttressing are the heart-pine corner posts and the palm logs.

And the log walls are resplendent. Some are vertical, stockade fashion, others horizontal, Lincoln Log–wise. But they are deployed creatively, either to maximize the use of shorter leftover log ends or because the workmen had style as well as skill. Style is evident in the palm-wood fireplace and the detailing on a palm log table upstairs. Sang told me that there had been more substantial palm furniture and the upstairs table persisted only because of its ponderous weight. On the north kitchen and dining side of the first floor Sang had the logs painted white to contend with the inherent darkness. The shorter log pieces remind her of Kellogg's Frosted Mini Wheats with their bulging contours and fibrous aspect. Where the logs meet each other at right angles they are not cut square but meticulously coped, resulting in some unexpectedly undulating patterns.

The only remnant twig-work is a plaque that Anderson had made to display the Seminole name for the structure. Roberson maintains that name is Talahloko. The National Register thinks it is Talahloka. Sang points out that the final character is constructed like the first *o,* not the two *a*'s. But a provocative nail hole just to the right probably suggested to others a missing descender on the last lowercase letter that went missing. Sang suspects it was a little brace. Adding to the confusion is an online English/Seminole dictionary that renders big cabbage as "ta-la thak-ko" and cabbage palm as "tah-lah-kul-kluk-ko" on page 12 and palmetto tree as "tala-ka-kulke" on page 20.[7] Old postcards agree with Sang, but one postcard suggests the home might be a typical Florida bungalow—an unlikely characterization for what may have been the only historic two-story cabbage palm log building ever created. But that is not to say it is the only palm log home.

The densest-known concentration of existing palm log edifices is in Myakka River State Park, where eight palm log structures built by the Civilian Conservation Corps between 1935 and 1941 not only still exist but remain in active use. Not only is this the public's best opportunity to see palm log construction—you can even stay in any one of five cabins. After eighty years of use (and abuse from hatchet-wielding vandals) the park restored most of the palm structures by elevating them and replacing bad logs. Annual flooding had compromised some of the lower logs, as had drips from air conditioners. Nearly four hundred replacement palm logs were harvested in the park from a designated 100-acre plot. They were dried for three months and then cut to fit the live-oak framing. The park's logs run horizontally, and they alternated orientation so that the slightly smaller diameter top end of one log rested on the slightly wider bottom end of the previous log. In keeping with the historic construction, they were wired to the frame with #5 copper wire and chinked with a mixture of tar and sawdust. The project superintendent, Jeff Conkey of DEC Contracting Group, Inc., told me his firm had never worked with palm log construction before but that it had been a fairly straightforward eight-month project. His main tips are to let the logs dry, keep them off the ground on a flat surface (when fresh and wet they will assume the shape of whatever they are stored on), and to wear gloves.

The postwar demand for housing encouraged innovation. In 1946, C. L. Baker, a stonemason in Port Richey, hit upon a new approach to cabbage palm construction. He sawed the trunks into 8-inch sections and then, being a mason, stacked them in mortar to form the walls of a house. Then he covered the exterior with cement, and cut grooves around each log and filled the grooves with black cement, creating the appearance of a stone wall comprised of circular stones. In the process he learned for himself how problematic sawing palm logs can be as the logs quickly dulled the blade.

Three years later, in 1949, a retired Miami physician, C. J. Caraker, tried another approach in St. Petersburg. He claimed to have optioned 4.5 million acres of Florida palm forest to support his entry into the low-cost housing market. Caraker set his logs vertically, leaving the exterior natural, but inside the logs were "buffed down to a velvety texture of variegated tans and browns with the joints between logs stripped with natural finish triangular cypress strips." He acknowledged the logs were rough on saw blades, requiring them to be resharpened every three hours, an outcome he attributed both to "innumerable long fibers" that clogged the teeth, and the fact that "the logs are said to be more than 80 per cent sand." The two-bedroom homes were to be sold for $7,500.

Caraker had been inspired by a home he had seen in the Crystal River area, one built by baseball Hall of Famer Dazzy Vance. Caraker had many claims for his promised palmetto log homes: they would be fireproof, ter-mite-proof, hurricane-resistant, "highly resistant to both heat and cold and have acoustical ability to absorb sound to a remarkable degree." But, per-haps unaware of the history of log fortifications, Caraker failed to mention that they would also be valuable in case of armed attack.[8]

I contacted the present owners of this now seventy-year-old home, and they confirmed that despite its contemporary stucco disguise, most of the home (except for the now closed-in former carport) is comprised of palm log walls. They knew they had a log home but were surprised to learn it had been a publicized attempt to create low-cost postwar housing. They had been told a far more exotic story when they moved in: the neighbor-hood explanation of the thick log walls was that it had been built for the local mafia, the logs being bullet-proof. Apparently, the clincher was a small window in the gable end of the roof that neighbors interpreted as a defended port where an imagined sniper could hide and pick off J. Edgar Hoover's St. Petersburg representatives when they arrived expecting to eas-ily capture the bad guys in their palm log fortress.

The best-known cabbage palm house may be one inspired by Tom Gas-kins, owner and developer of a classic Florida roadside attraction in off-the-beaten-path Palmdale, Florida. Known as Tom Gaskins' Cypress Kneeland, the operation lasted through the 1990s. In my opinion, Tom was the ulti-mate homegrown cracker entrepreneur. He started as a salesman for Gator Roach Killer and ended up with ten patents, including one for a turkey call. Those autographed turkey calls now sell for over one thousand dollars, if you can find one. His most intriguing patent was probably No. 2,069,580 for "Articles of Manufacture Made from Cypress Knees." Cypress knees are woody protrusions from the roots that emerge out of the swampy ground where cypress tend to grow. Tom didn't have a firm opinion about what

Advertisement for a low-cost cabbage palm house appeared in the *Tampa Bay Times,* July 16, 1962, 13.

functions the knees provided (nor, as it turns out, do researchers, according to an article in Harvard's botanical journal, *Arnoldia*),[9] but that didn't stop him from figuring out how to make a living by sawing them off and displaying or selling them. Tom's patented cypress knee lamps were made by Tom and his son peeling the skin off the knees and hollowing them out to accept the wiring. Despite having visiting Cypress Kneeland, I don't remember meeting Tom Sr., but once I heard about a palm log house associated with Tom, I resolved to meet his son.

Tom's son's mobile home burned down in 1972, and Tom Sr., who had always been intrigued by the possibilities of palm logs, urged him to consider a palm log home. At that point Tom Jr. had more time than money and decided to pursue the idea. Tom Sr. thought his son should bury the logs upright in the ground, but Tom Jr. had observed that palms making contact with the ground would wick water and rot, so he concluded that keeping them off the ground was going to be a key factor. Because he could visualize rainwater lingering on horizontal logs, he resolved his palm log exterior walls would be based on vertical orientation and he'd use two-by-fours and wallboard for any interior partitions.

Tom Jr. sketched out a 30' × 60' house on graph paper and decided he was going to need straight palm trunks as close to 9 inches in diameter as possible. He figured 9-foot logs would give him 8-foot ceilings with about a half foot to work with above and below. He found a seventeen-year-old helper, Dennis Lawrence, and located a stand of cabbage palms in a place called Boar Hammock, about 5 miles south of Palmdale. Boar Hammock was owned by Mr. Dilley, a former teacher of Tom Jr. Tom described the site as 84 acres that was 80 percent cabbage palms. Armed with a chainsaw, they started on the last day of March 1974. Tom agrees that palm logs are rough on saws and subscribes to the theory that it is the fiber that is especially problematic. Since they couldn't always park their pickup close to the trees, they would have to lug them out, and he still seems to retain a body memory of the fresh-cut logs being "as heavy as lead." Twelve logs were considered a good day's work, and they would stack them cross-wise on the bed of the pickup. Ever mindful of the potential for rot, Tom decided it would be wise to infuse each log with creosote from the center out, so once the logs were on the truck they would approach the cut end of each log with a 2-inch-diameter auger that had been welded to a 5.5-foot-long steel rod that was attached to a ¾ horsepower heavy-duty drill. Tom had spent some significant portion of his life boring cypress knees for his father's lamps, so he was able to keep the drill bit centered as it worked its way in. Of course, he had to back the drill out periodically to clear the accumulated waste. Since it was inevitably tempting to get just a few more inches in, the drill would bind sometimes. If it was in one of the logs on the top of the pile, the jammed drill bit would simply get the log spinning, but if it was a lower log with a lot of weight on it, then the log would remain fixed and proceed to start frenetically spinning the engine.

After they drilled the logs halfway through on one side of the truck, they'd go around and drill matching holes from the other side. Tom said he only missed meeting the previous hole on one log. Those drilled central cavities were a lot of work, but they facilitated three things: drying the logs faster, creating a space to eventually run wiring, and establishing an entry for the dose of creosote. The holes also simplified moving the logs around—they'd simply slip a 2-foot pipe partway in each end and lift the log with the temporary pipe handles.

Tom realized that placing two vertical cylindrical palm logs next to each other would result in very minimal contact between the two logs and require a lot of chinking, so he dreamed up an apparatus to minimally flatten a log's surface by creating a cradle of sorts that would enable him to slice off some of the curvature. Then each log would be flipped over and shaved again to yield a log with two opposing flattish sides (two adjacent flat sides for the corners). He said the younger logs went pretty quick, but he "paid a price" for tackling older, tougher logs.

When I chatted with Tom for two and a half hours in Venus, Florida, he stopped at one point to stress what he considered a crucial insight. One of his father's friends from Miami had noticed that the tough fibers that protrude from the trunks invariably point upward, so that convinced Tom to install each log upside down so the spiky fibers would divert water away from the log instead of directing water down the trunk. He's a firm believer in vertically oriented upside-down palm logs and can't really fathom why some people would install them horizontally or right side up.

While the logs were drying he created an elevated foundation of free-standing concrete piers and put in 5" × 10" heart-pine sills (from a sawmill in Olga) and 2" × 11" floor joists. He cut away part of the bottom of each log so it would sit on the sill with roughly a third of each log overhanging the outside of the sill. He then nailed two 20-penny nails through the overhang into the sill. At the top of the walls Tom created a plate for the rafters from two two-by-fours, leaving a 1-inch gap between them in case he ever need to douse the logs' central cavity with more creosote. Turns out he didn't. He hammered a bridge nail through each log to link it to the previous log.

Tom moved in on the last day of February 1975, forty-seven weeks and five days after they first started cutting in Boar Hammock. He said that year was the hardest he ever worked in his life. Two hundred seventy-six cabbage palms gave their lives to create an 1,800-square-foot home. He lived there for twenty-five years.

Unfortunately, Tom's father didn't own the land in Palmdale, and when the owners, Lykes Brothers, sold the land to the state, Tallahassee bureaucrats assumed the state would get Tom Jr.'s log home and Tom's parents' place. But, fortunately, the Glades County property appraiser had recorded all the Gaskins' structures as personal, not real, property, so Tom could keep the buildings, but he had to move them promptly. They moved the homes and outbuildings intact—visualize a 30-foot-wide house heading up the road for 18 miles, taking up both lanes and forcing all other vehicles onto the shoulder. His new place is still adjacent to Fisheating Creek. Over the years he has evolved through a variety of chinking techniques—roofing tar, neoprene rope, and, finally, liquid nails. When I met him he had lived in the new location for eighteen years, was anticipating his seventy-seventh birthday, and had begun wondering what would become of his forty-three-year-old home in the future. He couldn't bear the thought of some clueless future owner covering his unique dwelling with vinyl siding.

21

———•———

Cheesecake

The Most-Photographed Cabbage Palm

In 1960, Betty Frazee was a twenty-year-old model working at a natural Florida attraction known as Silver Springs. She had been discovered four years earlier when she was a drum majorette visiting the attraction from Ocala, and she started working there the next year as a "public relations officer." At eighteen she came in third in the Miss Florida contest. One of the prizes was a week's vacation in Sarasota with her mother as chaperone. That didn't prevent her from posing for illustrator Thornton Utz and pinup artist Gil Elvgren.[1] When I interviewed Betty, she had several scrapbooks filled with press clippings.

Betty was part of a symbiotic relationship between the national press and Florida tourism promoters. Photographers and publicists provided a torrent of free publicity stills to the media—visual chum that existed primarily to allow northeastern and midwestern readers to fantasize about vacationing in, or moving to, Florida.

Not coincidently, many of these images contained women in formfitting swimsuits: bathing beauties. Petite Betty (5 foot two inches), both blond and curvaceous, was no exception. Such publicity photos borrowed from both World War II pinups and glamour photography and were sometimes referred to as "cheesecake," presumably because they were seen as an indulgent, guilty pleasure, and double entendre captions frequently accompanied the photos. Pioneer underwater photographer Bruce Mozert defined cheesecake as "a good looking girl doing unusual things." That simple formulaic juxtaposition enabled Mozert to get publicity stills for

Florida's oldest nature-based attraction, Silver Springs, placed in not only male-oriented magazines such as *True, Argosy, Florida Outdoors,* and *Skin Diver,* but also high-profile, general-audience magazines such as *Life.*[2] Betty Frazee was one of his favored models and obliged Mozert by drinking Cokes underwater (an early product placement), as well as "talking to fish." But some of the most iconic cheesecake shots ever taken at Silver Springs weren't taken underwater.

As far as Google and eBay know, there is only one genuine Silver Springs TV Tray. Its clattering pressed-tin surface perched above its precariously scrawny wire legs features a dry-land photo taken by Mozert, and, as you might expect, it also features a cabbage palm. A brunette Silver Springs model poses in a modest yellow two-piece swimsuit standing on a sidewalk in white heels and appears to be leaning on a cabbage palm trunk that has taken a parabolic dive toward the Silver River. The famous glass-bottomed boats can be seen in the distance. The strong vertical format of this half of the image destined it to grace untold numbers of Silver Springs brochures. When one unfolds the brochure, the palm canopy appears suspended over the river.

That particular cabbage palm has been called the "lucky palm," the "wishing palm," and the "horseshoe palm," and it became famous due to the intersection of two phenomena: (1) it is very distinctly contorted, and (2) it is growing on the banks of the Silver River quite close to the headspring. As a result, it has probably been a noteworthy natural feature of the tourist attraction since the equally famous glass-bottomed boats first slid across the waters in the late 1800s.

No one knows why this palm has grown in such a droopy, loopy fashion, but the theory that enables it to do so is generally understood. Imagine a team of bricklayers building a cylindrical chimney. If the workers on one side add more bricks, or add larger bricks, or add bricks faster, the chimney will start to veer to the other side. A palm trunk is a cylinder made of cells, and if cells are increased disproportionately on one side, the tree will bend away from that side. That all seems quite understandable when a palm finds itself partially shaded and takes evasive action to seek more sunlight. A plant's ability to grow toward light is known as a positive phototropism, and it is one of two ways plants cope with being shaded. Plants with multiple shoots or twigs can appear to grow toward light when the shaded twigs decline and die and the sunnier shoots proliferate. But a

The Silver Springs attraction brochure featured a bathing beauty and the contorted lucky palm. Author's collection.

cabbage palm with only one growing bud does not have that luxury. The entire plant must do what it can to seek sunlight. Yet this palm's wandering ways are inexplicable, and its erratic novelty and high-profile location have made it the most-photographed cabbage palm in the world—appearing in both publicity stills and tourist snapshots for decades.

Comparable palms can be found elsewhere. There's another palm at Silver Springs called the "corkscrew palm," and the competing attraction Cypress Gardens boasted another "wishing palm"—a nearly horizontal queen palm that provided a perch for full-skirted southern belles to rest

Betty Frazee perched on the contorted lucky palm at Silver Springs. Postcard. Courtesy of the State Archives of Florida.

upon. In reality it is not hard to find distorted cabbage palms in any palm hammock, but none has had the exposure of the horseshoe palm at Silver Springs.

Yet the TV tray/brochure image is not the most iconic and unusual image taken at Silver Springs. That distinction belongs to a photo of Betty actually perched on the tree inches above the river. She holds a beach ball and appears to be comfortable, if not actually languishing, in the curves of the palm—truly a good-looking girl doing something unusual. The horizontal format became a popular postcard that simultaneously featured an exotic and contorted tropical palm, a beckoning bathing beauty, and the Silver Springs attraction.

One has to ask how a palm trunk the diameter of a dinner plate can support even a petite model without breaking. Other postcards reveal a post emerging from the spring run that is supporting the palm. Many historic publicity shots of the palm use creative angles and cropping to conceal this crutch.

At some point the horseshoe palm was hoisted up and given its own deck that extends out over the spring run. The palm seems safer and is far more accessible, and the wooden deck has become a popular spot for what was once called a "Kodak moment."

Silver Springs, however, is wobbling. It started as a simple natural attraction in the 1860s and experienced a breakthrough by utilizing glass-bottomed boats to access the wonder of the springs. But what had been a natural attraction evolved into a complex of diverse recreational opportunities that at various times included a Seminole Indian village, a rattlesnake-venom-milking Reptile Institute, Six Gun Territory, wild (escaped) monkeys, an Early American Museum, Jeep Safari, the Prince of Peace Park, an albino alligator exhibit, an antique and race car museum, Paradise Park, and a zoo with a lonely giraffe. "Paradise Park" was a Jim Crow–era portion of Silver Springs ("for colored people") that from 1949 to 1969 allowed the black community limited access to the attraction.

Today the giraffe has died, the water is contaminated with nutrients, and the flow has declined dramatically. Most of the ancillary attractions are gone. The fates of both the natural springs and the attraction are being debated. But the lucky/horseshoe/wishing palm remains, its tortuous convolutions mimicking the attraction's erratic path.

22

---·•·---

Fine Artists Contemplate
the Cabbage Palm

The Hudson River school landscape painters were associated with Roman-ticism and favored New England geography, but several of the Hudson River artists found their way to Florida. These include Thomas Moran, Martin Johnson Heade, Albert Bierstadt, and Hermann Herzog. Another renowned painter with a Florida connection is portraitist John Singer Sargent, who, in addition to the notorious formal *Portrait of Madame X,* created a memorable portrait of six muddy alligators while visiting John D. Rockefeller in Ormond Beach.[1] Sargent also found time to paint at least three watercolors: one of nearly a dozen tall cabbage palms[2] and two that appear to be young cabbage palms.[3] In fact, it seems as though most painters visiting Florida have a hard time not painting some cabbage palms. So I've come to enjoy looking for cabbage palms indoors in museums as well as outdoors.

The most accessible collection of Florida fine art paintings exists at the Cici and Hyatt Brown Museum of Art at the Museum of Arts and Sciences in Ormond Beach. The four hundred paintings exhibited at any one time represent about an eighth of the total, so the paintings are rotated. When I was there in 2016, there were hundreds of landscapes, many depicting cabbage palms. There were several gorgeous Herzogs and a N. C. Wyeth ink wash (*The Landing Party*) that captured the palm's spirit with gesture, and a watercolor sketch by Eleanor Matthews (Eatonville, 1950) that also captured cabbage palm character, even if it wasn't botanically accurate. I didn't see any works by my favorite palm portraitist, Winslow Homer.

For my money, Homer is the most masterful painter of palms, which he typically rendered in watercolor. His coconut palms are simple and accurate (and his blustery stop-motion scenes—*A 'Norther,' Key West*—foreshadowed their use as obligatory hurricane wind speed indicators by your local action news team).[4] To his credit, and my allegiance, Homer was equally adept at capturing cabbage palms, a species whose visual essence seems to elude most artists.

A particularly striking Homer is housed at Jacksonville's Cummer Art Museum—*The White Rowboat, St. Johns River*, painted in 1890. Out of a tidal marsh, four healthy cabbage palms tower over a dying oak that is draped with Spanish moss. Three men in their eponymous white boat pass in the foreground. According to the Cummer Museum, "The towering, indigenous palm trees stand as testaments to the dominance of nature over man."[5] Perhaps. Or maybe Homer just liked including elements other than water and vegetation.

Homer painted numerous scenes in Florida, and there are other Homers in Jacksonville, but they are not accessible to the general public—they are part of an extensive private collection in a riverfront home. Fortunately, the collectors, Sam and Robbie Vickers, are gracious docents. The walls of the Vickerses' labyrinthine hallways and rooms are covered exclusively, sometimes floor to ceiling, with Florida art, and while many are landscapes and portraits, there are also juke joints, still lifes, and original *Saturday Evening Post* covers.

An appreciative comment regarding any one of the paintings results in either Sam or Robbie embarking on a story. Frequently their comments reflect their researched knowledge of the lives of the artists, how and why they came to paint in Florida. At other times they interpret the painting, or relate how they acquired the work, or reveal their personal familiarity with the locales. It becomes clear that beyond the artist's experience of Florida, a tour of the Vickerses' home is a tour of a couple and their relationship with both artists and the Florida the artists captured.

To my hopelessly biased eye, a significant percentage of the paintings seem to feature cabbage palms, and I fish about, hoping one of the Vickerses would start gushing about how nothing says Florida in a landscape painting quite like a few well-placed cabbage palms. But Sam responds with a pragmatic observation: "Tough to avoid 'em."

Winslow Homer (American, 1836–1910), *The White Rowboat, St. Johns River,* 1890, watercolor on paper, 14 × 20 in. Bequest of Ninah M. H. Cummer, C.0.154.1.

Stymied in my efforts to elicit an unqualified endorsement of the premier role of cabbage palms in Florida landscape paintings, I accept the fact that it was too much to hope that the Vickerses would openly share my obsession. But the collection speaks for itself, and despite their modulated response, I had to conclude Sam and Robbie, subconsciously at least, are somewhat smitten by our state tree. Or, at a minimum, they are smitten with artists who were smitten with our state tree.

Despite my thinly veiled quest to find the best palm-dominated landscapes, I found I was unable to remain focused and succumbed to the diversity and quality of their collection. I couldn't help but linger at N. C. Wyeth's glowing illustration from *The Yearling* of Jody and Flag, Frederic Remington's classic engraving of *Cracker Cowboys of Florida,* and an exquisitely detailed 1846 map of Florida that is about as tall as I am.

Beyond such gems, there were a lot of native palm paintings. I mean, dozens. William Lamb Picknell, Adolph Robert Shulz, and Laura Woodward all captured some great palms, yet it was the work of two artists that started consistently drawing my eye—Charles Robert Knight and

Hermann Herzog. Their work stands out not simply because so many of their paintings contain cabbage palms but because of their ability to capture sunlight. My assessment of Herzog's Florida works (which do seem to favor cabbage palms) reveals that he sometimes stumbles on accurately capturing the frond's anatomy—in many paintings the fronds are too feathery, the canopies too open and anemic. So I'd argue that Knight is the better of the two when it comes to rendering palmettos.

These two artists are quite nearly opposites. Herzog, born in Germany, was strictly a fine artist and ended up financially independent. Brooklyn-born Knight was financially vexed despite producing conventional oil paintings, drawings, sculptures, watercolors, large murals, children's book illustrations, and stained glass window designs. Although legally blind, Knight was an excellent draughtsman, and his work appeared in *McClure's, National Geographic,* and *Century Illustrated Monthly Magazine.*

Herzog painted more than three hundred scenes of Florida.[6] At least eight of his works are in the Cummer collection, and one is a palm-featuring Florida landscape.[7] Knight has none in the Cummer collection, and I didn't see any on display at the Brown Museum. A Google image search yields dozens of Knight paintings, yet none depicts a nineteenth- or twentieth-century Florida landscape. What gives?

What sets Knight apart from other fine artists is not so much the fact that his landscapes appear in some of the best-known museums in Los Angeles, Chicago, and New York, but that the museums are natural history museums. For Knight was far more than a landscape painter. He pioneered the depiction of prehistoric animals, developed an anatomical approach to rendering wildlife, and visualized what became a popular exhibit at the 1964 World's Fair.

So is the Vickerses' collection the only place to see a Knight cabbage palm? Well, there's a saber-tooth and mammoth mural in the Sebring, Florida, public library. Truthfully, they are not his best cabbage palms, although it is a Florida landscape. The better answer is that if you or your children ever went through a dinosaur-obsessive phase, you've probably already seen a Knight cabbage palm painting. They appear in the background of a large mural painted in 1930 for the Field Museum in Chicago. In the foreground a menacing triceratops threatens an equally menacing *Tyrannosaurus rex.* This is quintessential Knight—prehistoric animals not stiffly posed but interacting in a realistic landscape that contains scattered clumps of palms that could be somewhere in a Florida prairie (aside from the fact that the Florida peninsula was submerged when dinosaurs stalked the Earth). This confrontation "has been borrowed, copied, and reproduced countless times in textbooks, comic books, and movies."[8] And that's not all. Cabbage palms also appear in his painting of Uintatheriums.

Knight dreamt of a park in Florida that would feature giant, life-sized models of dinosaurs. His vision was not without irony since Florida is the least dinosaury state. The facility was to be called Dinoland, and Knight was unable to persuade anyone to build it. After Knight's death in 1953, a younger collaborator and natural history artist, Louis Paul Jonas, pursued the idea, and his efforts eventually resulted in Sinclair Oil's 1964 New York World's Fair exhibit, called Dinoland.[9]

After Homer, Knight, and Herzog, the masters, cabbage palm rendition falls off fairly rapidly, in my opinion. There's good reason for that—cabbage palms are very hard to capture with anything other than a camera. I believe that has to do with the fronds. When painting a landscape with dicot trees, no one bothers, or attempts, to paint every leaf. But with a palm, it's hard not to deal with each leaf. Coconut and queen palms, the pinnate palms, have feathery plume-like leaves. Some of the palmate palms have leaves that are more constrained to a single plane. But cabbage palm leaves are costapalmate, which means only the earliest leaves when the palm is emerging from the ground, are flat. Therefore, a mature cabbage palm leaf cannot be rendered as a flat fan. To complicate matters, the individual fronds combine numerous relatively straight elements as well as conspicuous droopy curves. So artists tend to veer toward two extremes: too feathery/fluffy, or too spiky. Herzog leans toward feathery; Knight, spikey.

We think of John James Audubon as a bird portraitist, but many of his images include backgrounds, and the Louisiana (now tricolor) heron features many cabbage palms in the distance. The background wasn't provided by Audubon himself but rather by an assistant, George Lehman,[10] "a lithographer, engraver, ornamental painter, and aquatintist" who accompanied Audubon to Florida in 1831 and 1832.[11] It was while looking at Lehman's decidedly feathery palms that I noticed a number of dead, presumably live oak trees. I know from personal experience that it is sometimes tempting to simply draw the trunks of trees without the leaves as an expedience, so it is possible Lehman was just short-cutting. But it got me wondering if he had faithfully reproduced a particular waterfront setting and, if so,

Did this 1904 painting inadvertently depict rising sea level in the Homosassa area? Winslow Homer (1836–1910), *The Turkey Buzzard,* watercolor over graphite on paper, Worcester Art Museum Massachusetts, USA/Bridgeman Images.

what killed those trees. Aside from old age, trees can be killed by lightning, disease, and changing environmental conditions.

That brought me back to Homer's *The White Rowboat* in the Cummer Museum. These aren't Homer's best palms—kind of green fuzzballs without the curved components seen in some of his works from the Big Bend, but that's not the point. There are only five trees in that painting, four healthy mature cabbage palms and a dead live oak draped in Spanish moss (there is a smaller dead tree or shrub on the right). Not only are there no living live oaks or other dicot trees, but there are also no young cabbage palms. The absence of young palms allows the tall palms to be seen in all their elegant linear simplicity. But the absence of young trees also means that decades earlier conditions ceased to be favorable for the germination and establishment of palms. One likely explanation is that increasing amounts of salt or brackish water due to rising sea level prevented subsequent generations of palms from becoming established. So the question is, did Winslow Homer, looking for great subject matter on the St. Johns

River, inadvertently capture the earliest image of the effects of rising sea level?

I started paging through coffee table books that contained Homer's Big Bend works. It is not clear where *In a Florida Jungle* (1885) was painted, but it depicts a patch of saw palmetto viewed from the water, with an alligator and eroded cabbage palm in the foreground. In the background a cypress tree on the right is balanced by a stand of tall cabbage palms on the left. No young cabbage palms are visible, but since it may not be coastal, it is a tangent. Two Big Bend paintings seemed ambiguous as far as capturing rising seas. *Homosassa Jungle in Florida* (1904) had a dead oak but also a masterfully rendered youngish palm with its boots on. And *The Shell Heap* (1904) had many tall, white-trunked palms but also two shorter, booted palms. Then I came upon *The Turkey Buzzard* (1904), also believed to be painted in the Homosassa area. More than a dozen bare-trunked palms form a hammock island with a slash of straw-colored marsh or shoreline. The requisite turkey buzzard (turkey vulture) soars in the upper right, dangerously close to drifting out of the frame. Here, as in *The White Rowboat,* Homer clearly shows us a dying palm hammock, suggesting that the decline of coastal forests associated with rising sea level in the Big Bend region was well under way more than a century ago.

23

———•———

Toby Kwimper

A Cinematic Experiment in Ecological Succession

As historian Gary Mormino is fond of pointing out, there is something about "the Florida Dream" that resonates in a way most other states can't match. There is little talk, for instance, of the Nebraska Dream or the Rhode Island Dream. But Florida is a dream state, and dream states, it seems, are the places where one can go to start over.

I guess it was the lure of this dream state that inspired Garden State resident Pop Kwimper to pack up his fifty-dollar car and leave "Cranberry County, just over the state line," taking his family to Florida. In addition to Pop, the cobbled-together group consisted of his naïf back-from-the-army son, Toby, the fourth-cousin twin boys Teddy and Eddy, nineteen-year-old Holly Jones, who was just the twins' babysitter until her parents died in a car wreck, and Ariadne, a three-year-old orphan of unknown provenance.

Pop's girl-shy son, Toby, was making $63.80 a month on total disability from the army. The twins were getting aid for dependent children, and Pop was "on relief." When their jalopy ran out of gas next to a sandy beach graced with a few palm trees, they decided to homestead—proceeding under the imaginative impression that if they put up a roof and stayed six months, it would be theirs.

The beach was "filled land" created by the government when tons of sand were dumped on top of a tidal marsh. But before I get much further into this tale you need to know two things: first, that these factoids are part of the plotline of a 1962 MGM/United Artists film called *Follow That*

Dream and, second, that the Toby Kwimper character was played by Elvis Presley, starring in his ninth film.[1]

The sand was dumped not by the government, as the plot would lead you to believe, but by the movie production company, and they dumped it on a tidal marsh next to Pumpkin (or Punkin) Key on the northeast side of the Bird Creek bridge on State Road 40 leading out of Yankeetown, Florida. The Florida Development Commission threw in eight thousand dollars for "clearing, filling, and dredging the area to suit the script."[2]

The addition of a number of transplanted, droopy coconut and cabbage palms (they had to paint them green partway through filming)[3] lent a tropical flavor to the sandy expanse. Early on in the film the mischievous twins attempt to dislodge coconuts by throwing rocks (at three cabbage palms!) while Holly brings Ariadne some water. Toby (Elvis) is seen manipulating a palmetto frond before climbing up on a rustic cypress pole structure to thatch a roof.

About nine minutes into the movie, Toby descends from his roof work and absentmindedly fingers a palmetto frond while he attempts to explain to Holly (who considers herself "pretty well built") why he is immune to her charms. Cue Elvis song: "I'm Not the Marrying Kind." Sample lyric: "Don't kiss me, don't claw me, don't pet me, don't paw me, I won't leave my freedom behind."

Several websites that attempt to rank Elvis's thirty-one films include *Follow That Dream* in the top ten, and at least three rank it as his best film.[4] In the course of the movie a guileless Toby secures a bank loan without collateral, runs off a Mafia gambling operation, resists a seductive, predatory social worker, wins a court case, (and succumbs to Holly). But those plot points obscure the fact that notwithstanding the 1991 Julie Dash film *Daughters of the Dust* and Burl Ives's role as Cottonmouth in *Wind across the Everglades, Follow That Dream* is clearly the pinnacle of the cabbage palm's cinematic career and would be worth including in this book, if only because of Elvis's absent-minded frond handling while he chats with Holly and the palm-ridden setting. But there is more to the story.

In a landscape where a few inches of elevation can result in a completely different plant community, much of the filled tidal marsh set was destined to turn into some other habitat as seeds colonized the fill. The Mirisch Production company (which also produced *West Side Story* and *The Great*

Toby Kwimper (played by Elvis Presley) climbs a cypress pole structure with a cabbage palm frond to thatch a roof. FOLLOW THAT DREAM © 1961 METRO-GOLDWYN-MAYER INC. Courtesy of MGM Media Licensing. Elvis Presley™; Rights of Publicity and Persona Rights: Elvis Presley Enterprises, LLC. elvis.com.

Escape) had inadvertently created an experiment in ecological plant succession: What would happen next if you covered a black needle rush marsh with sand and left it alone?

If you turn off of Highway 19 in Inglis to head out to the Bird Creek Bridge, first you'll pass the historic marker acknowledging the filming, and then pass through Crackertown, a defiant answer to Yankeetown, which lies a little way down the road.

Yankeetown had been founded in 1923 by some Yankees from Gary, Indiana. The Knottses, perhaps understandably, were inclined to call the place Knotts. A Knotts descendent, Tom Knotts, has written a modest history of the area: *See Yankeetown.* Tom dedicated his book to A. F. Knotts, "attorney, civil engineer, scientist, ethnologist, historian, conservationist, and founder of Yankeetown."[5] Yankeetown is home to several cabbage palm log cabins, but if most people have heard of Yankeetown at all, they know it as a fishing locale on the Withlacoochee River, home of the Izaak Walton Lodge, or the place where they filmed that Elvis movie.

As far as I can tell, Yankeetown and nearby Homosassa represent the general epicenter of cabbage palm log construction. There are the CCC-built cabins at Myakka River State Park and the two-story Talahloka residence in Ormond Beach, a barn in Inverness, and other scattered log homes, but palm log construction was apparently fairly common in this section of the coast. There is a beautiful hybrid construction home on Riverside in Yankeetown that consists of vertical palm logs with opposing grooved sides that receive pecky cypress boards—a board-and-batten variant. A nearby home on Riverside features palm log porch supports, as does the local real estate office. One or two workmen willing to learn how to build with palm logs would have had all they could use, basically free for the taking.

There are few residences along Highway 40, which is now dubbed the Follow That Dream Parkway, and the major function of this dead end road seems to be to get people to the terminal boat ramp. Along the way one passes through an excruciatingly scenic landscape of oak and palm hammocks, tidal marsh, and tidal creeks. Just don't look to the south because the cooling towers of the now-closed Crystal River Nuclear Power Plant four and a quarter miles away tend to dominate the horizon. Less than a mile from the end of the road is the Bird Creek Bridge, and, since there is no historic marker at the actual movie location, most boat trailers and curious tourists blow right by, the ever-expanding vistas communicating that the end is near.

When I first saw the movie I was nearly tempted to hold out hope that some of the transplanted palms had survived. I didn't know about the green paint. But the palms were clearly declining in the movie, and you can't transplant palm trees with all their fronds intact and expect them to survive, especially in a salt marsh and without copious fresh irrigation water. Some say the crew did what it could to remove most of the sand once shooting was completed, but there was no feasible way to remove all the sand and get back to marsh plants at their original elevation. "Whatever the moviemakers left behind, like the artificial white-sand beach, was reclaimed by nature."[6] Aerial photography taken in 1963 shows more than 3 acres of what appears to be mostly bare sand. Hence the inadvertent experiment in plant colonization.

So what's there now, six decades since the film crew left? Either the northern half of the site was covered with just enough sand, a frosting, to create a cinematic beach in the distance, or sea level has risen enough to allow tidal marsh to prevail. That area is once again a black needle rush marsh, a handsome expanse of dark rushes that come to decidedly sharp points, sharper than knitting needles.

The area closer to Highway 40, where the action in the movie was shot, was buried in at least 2.5 feet of sand. That area, the former Kwimper homestead beach, has become a new hammock. This happens to be a hammock dominated by cabbage palms and southern red cedar, which is technically a juniper, and a reliable component of the Big Bend coastal hammocks. There are a few live oaks, one pine tree, and some Florida privet, but it is the cabbage palms, some with 25 feet of trunk, that define the parklike locale. The groundcover is bahia grass, so with a little mowing and a little pruning the site would make a nice shady place to picnic. The property is owned by the Felburn Foundation, a Silver Spring, Maryland, nonprofit. Perhaps someday a memorial plaque will be placed on the northeast side of the Bird Creek Bridge, commemorating the Inadvertent Toby Kwimper Homestead Memorial Plant Succession Experiment Hammock Park and Wayside Rest.

24

———•———

Elegance Doesn't Come Easy

The South Carolina State Flag

I'd never heard of the blog *Thrillist,* but *Thrillist* went to the trouble of ranking the fifty state flags, and, guess what, they ranked South Carolina's as the number-one state flag.[1] The *Houston Press* also took a shot at ranking state flags and found South Carolina's to be fourth-best, behind those of New Mexico, Arizona, and Colorado.[2] So, best flag east of the Mississippi.

But for authoritative determinations on such matters, one should probably skip the blogs and turn to the paid members of the North American Vexillological Association. It surveyed its members in 2001, and the South Carolina flag ended up tenth (Florida's flag ranked thirty-fourth).[3]

Whether the South Carolina flag design is the best or just somewhere in the top ten, it is clearly one of the most popular state flag designs among both internet users and vexillologists. The simple, elegant design suggests there was a simple, elegant process that led to the design. Not so.

Certainly Colonel Moultrie used a straightforward approach to the provincial flag used at Fort Sullivan. He borrowed the background blue from the troops' uniforms and added a simple crescent in what we think of as the upper left corner, but which in heraldry is referred to as the upper right (dexter) since shields were interpreted from the perspective of the owner. Moultrie was quoted as saying, "I had a large blue flag made with a cresent [*sic*] in the dexter corner, to be in uniform with the troops."[4] One assumes that meant the horns or points of the crescent faced upward, reflecting the horns-up emblem on the troops' caps; however, contemporaneous illustrations of the battle suggest two other orientations for the horns.[5]

Exactly what the upturned crescent on the caps represented at the time is another matter. A heraldic symbol (denoting a second son?), a crescent moon, a vestigial gorget, or some mashup of all three are possibilities.[6] An enduring shape, it benefitted from appearing early and is one of two symbols on the current South Carolina flag, where it is routinely, and no doubt inappropriately, interpreted to be a waxing (not waning) crescent moon.

The unexpected success of the palmetto that comprised palm log Fort Moultrie quickly found itself incorporated in the state seal, known as the Great Seal.[7] In the decades that followed, the palmetto kept reappearing on flags, usually symbolizing South Carolinian resistance, whether it was in regard to the Mexican War or nullification. (Nullification was a doctrine advanced by South Carolina in 1832 that the state could nullify federal laws, in this case, tariffs.) As South Carolina began moving toward secession, competing flags or banners featuring palmettos began appearing, which illustrated the reality that since there was no officially designated flag, anyone with an idea and access to fabric and a seamstress could produce one. After the Ordinance of Secession passed on December 20, 1860, the legislature set about, among other things, adopting an official design for a state flag. But the need for flags and the ability of people to produce their own wouldn't wait for the legislature.

Flag images today may be either sewn or printed. Printed flags can contain intricate detail, such as that found in Florida's state seal, not easily possible with sewing. Early American flags were either sewn or painted, or both. Two sewn Civil War–era South Carolina flags embody the challenges inherent in depicting a palmetto solely with fabric. These are the Citadel's "Big Red" Flag[8] and the Palmetto Guard flag.

"Big Red" entered the history books on January 9, 1861, when cadets from the Citadel (South Carolina Military Academy) fired on a ship attempting to resupply Fort Sumter. They were flying a flag that featured an appliquéd white palmetto and crescent on a "blood-red" field. A version of the flag eventually became the Citadel's spirit flag. Remarkably, in 2007 a flag matching the description of "Big Red" was located in an Iowa history museum. Research conducted by the Citadel concluded it was the actual flag flown on Morris Island in 1861.[9]

The "Big Red" flag is believed to have been made by Charleston flagmaker Hugh Vincent.[10] Vincent's challenge was to portray a palmetto, which is, if nothing else, a vertical design element, on a horizontal field.

This design constraint forced the tree to be squat. Another challenge was how to depict individual leaves in an effort to make the palm canopy look like something other than a roundish meatball-on-a-post lump. Vincent's symmetrical graphic solution with adroit use of negative space is reminiscent of traditional floral Hawaiian quilts, although the palmetto is bilaterally, not radially, symmetrical. It can be argued that the result does not look very much like a palmetto, but, to its credit, it also does not look like any tree other than a palmetto. It may, after a century and a half, be the best graphic representation of a palmetto to ever appear on a flag. After spending a century and a half in Iowa, "Big Red" was returned to the Citadel in 2010.[11]

The Palmetto Guard flag (aka the Lone Star and Palmetto Flag) was the first flag raised by Confederates over Fort Sumter at the start of the Civil War.[12] It consists of an appliquéd palmetto and a painted red star. It first gained fame when flown in New York harbor in November 1860 and was subsequently returned to South Carolina, where it came into the possession of Private John S. Bird, who used it on April 14, 1861, as the unofficial company flag of the Palmetto Guard.[13] While bold and dramatic, the placement of the negative space suggesting the area between leaves is not as convincing as on the "Big Red" flag.

Meanwhile, less than two weeks after the "Big Red" incident, on January 21, a joint legislative committee presented a resolution mandating "a National Flag or Ensign of South Carolina shall be white with green palmetto tree upright thereon; and the union blue, with a white increscent." In heraldry "increscent" dictates the horns pointing left. The upper left corner of a flag, known as the canton or union, is a place of honor because that is the portion that is visible when the flag is hanging limp.[14] The white increscent in the blue union was meant to be reminiscent of the flag flown at Fort Moultrie in 1776. Robert Barnwell Rhett Jr., legislator and editor of the *Charleston Mercury,* countered with a proposal that "the National Flag or Ensign of South Carolina shall be blue with a white palmetto tree upright thereon, and a white crescent in the upper corner." Since Rhett didn't specify which direction the crescent was to face, he can be credited with designing the current flag. But his "simple and beautiful flag" was not immediately embraced by the legislature.[15]

One legislator favored a green palmetto, but that was opposed based on the premise that green fabric would fade to a "dirty yellow." The Senate

The "Big Red" flag may be the most adroit depiction of a palmetto ever to appear on a flag. ™The Citadel Alumni Association.

had other ideas, including a flag with a coat of arms, and a flag similar to the one proposed by the joint committee except that it would be a red flag instead of white with some brown added. Despite controversy, that design squeaked by on a 30 to 28 vote. Later that day, a motion to reconsider led to going back to a white tree on a blue flag, and that recommendation was forwarded to a conference committee. The committee must have been feeling either creative or rebellious for on January 26 they came back with "a golden palmetto upright upon a white oval in the centre thereof" on a blue flag. The white increscent persisted.[16]

A week had passed since the joint committee first suggested a flag design. Someone using the pseudonym "Secession" wrote to the *Charleston Daily Courier* arguing that either gold gilt or yellow fabric would prove unsatisfactory for practical reasons related to weathering. On January 28 the House respectfully requested the concurrence of the Senate in altering the flag "to wit: dispense with the white medallion and golden palmetto, and in their place to insert a white palmetto." There was no mention of a crescent or increscent. The Senate concurred. Rhett, whose version had prevailed, used his paper to clarify: "It now consists of a blue field with a white palmetto in the middle, upright. The white crescent in the upper flag staff corner remains as before, the horns pointing upward. This may be regarded as final."[17]

That fact that Rhett viewed his description as "final" didn't resolve the design. Rhett as editor was clear that the horns of the crescent were to point upward (Moultrie-cap-style), whereas the conference committee design stipulated that the horns face the flagstaff.

Throughout all the negotiations and color changes, the palmetto, whether green, white, or golden, was deemed essential. For eighty-five years the palmetto tree had represented South Carolinian independence and defiance—one toast was, "The Palmetto—Ominous to oppressors!"[18]

In 1899, reflecting upon the Civil War, Representative Thomas Bacot of Charleston suggested that the ground be changed from blue to royal purple. Bacot argued that the change represented the blue flag soaked "in the red blood of the gallant South Carolinians." The measure failed.[19]

In 1913 the *Orangeburg Times and Democrat* reported that veteran survivors of the Battle of Gettysburg gathered on the occasion of the fiftieth anniversary of the battle. When the South Carolina division arrived at the camping ground, they were delighted and gratified to see a splendid palmetto tree that had been shipped from James Island to welcome back the veterans of the Palmetto State. It was a curiosity, and people requesting souvenirs were given streamers from the fronds. Eventually the trunk was cut in two, and "by the time the reunion was over not much more than the roots was left." "It was the Yankee soldiers who were most interested." "Scores of them have pieces of it which they will prize as souvenirs of the happy meeting of former foes." The article went on to relate how W. A. Clark of Columbia, who had sent the palmetto, decorated five "handsome, fine looking old ladies" who "as young girls were nurses in the federal army and who volunteered to nurse the wounded Confederates after Gen. Lee's surrender."[20]

By 1910 concern had grown regarding various details of the flag—not only the orientation of the crescent but also the color of the flag and size and shape of the palmetto. On February 26, the General Assembly approved Act 406, which, among other matters, reaffirmed the actions of the 1861 General Assembly, declaring that the flag "shall consist of blue with white crescent in the upper flagstaff corner, and white palmetto tree in the centre." By deleting two letters, the left-pointing increscent simply became a crescent, liberated to point in virtually any direction.[21]

Alexander Salley was the secretary of the Historical Commission, the body empowered by the General Assembly to resolve these issues. The color was the easy part—"neither flashy or dull—not so dark as to appear black from a distance, but not royal blue either."[22] The crescent was harder.

Salley was contending with two contradictory historical precedents: the upward-pointing horns featured on the caps of Moultrie's troops and the 1861 General Assembly's heraldic increscent with the horns pointing to the left, facing the flagstaff. Salley, apparently on his own authority, rotated the crescent halfway between the two options, thus deliberately or accidentally creating the familiar crescent moon orientation, the moon that many believe the crescent represents. Whether Salley was exercising a personal aesthetic preference, negotiating a split-the-difference compromise, or was honoring some earlier flag example he found appealing, is not known. Whatever his reasoning, Alexander Salley was the man who put the moon on the flag. In 1972, astronaut Charles M. Duke Jr., of Lancaster, put the flag on the moon. Not figuratively, he carried several miniature South Carolina flags with him on Apollo 16.

As far as what the palmetto was or is supposed to look like, there was some correspondence between Salley and Ellen Heyward Jervey, "an educated Charleston woman with an artistic flair,"[23] and Charles S. Doggett, director of Clemson's Textile Department, where the flags were made. And while the exchanges resulted in flags being made, apparently no written descriptions survived or were ever adopted setting either specifications or parameters for the palmetto tree.

As a result, despite Salley's work to resolve what the state flag should look like, there is apparently no definitive, officially adopted set of rules regarding the appearance of the flag save the original mandate that it be "blue with a white palmetto tree upright thereon, and a white crescent in the upper corner." That simple description allows wide latitude in what the flag actually looks like. That stands in sharp contrast with the exacting specifications for the U.S. flag, which includes ten specific proportional measurements (since the flag can be made in many sizes) and definitively stipulated colors (Blue PMS 282 and Red PMS 193, aka Old Glory Blue and Old Glory Red).[24] Florida's state flag also stipulates proportions: "The seal of the state, of diameter one half the hoist, in the center of a white ground. Red bars in width one fifth the hoist extending from each corner toward the center, to the outer rim of the seal."[25]

One consequence of this relaxed approach to South Carolina flag design is that there are currently several different versions of the state flag being sold by competing commercial manufacturers. The South Carolina State House Student Connection website features a version with a rambling palmetto sold by Flags Importer.[26] It is "printed on one side all the way through the fabric," which suggests that when seen from the obverse side, the crescent would be on the right. This version has the most irregular palm canopy outline of any of the flags currently being sold. The lower portion of the trunk is featureless with straight sides and leads to a section of bootjacks.[27]

The most common, possibly the most popular version, has a palmetto emerging from a small patch of ground with a tapered trunk and diagonal leaf scars—neither of which are characteristic of palmettos. No bootjacks are present. The canopy consists of eleven or twelve protrusions that have been compared to feathered wings or gloved hands.[28] This flag is sold, and presumably manufactured, by Annin Flagmakers, the "Oldest and Largest Flag Manufacturer in the United States—Since 1847."[29]

I understand that, as a Floridian, I have no appropriate role in characterizing South Carolina's flag—I am merely commenting on the rendering of the palmetto, and, in my opinion, the flag currently available with the more accurately rendered palmetto would be the version sold by US Flagstore.[30]

Other versions exist, and I found it hard to believe that *virtually any* upright white palmetto would do, and therefore I initiated correspondence with the South Carolina Department of Archives and History. They sent me a packet dated June 8, 2016, with an explanatory paragraph. The first sentence reads: "It seems that you are correct in your hypothesis that there is no officially adopted prescription for how the Palmetto tree is to appear on the S.C. State Flag." After citing the 1861 and 1910 legislation, the last sentence reads: "No other specifications have been officially adopted."[31]

Coasters, tote bags, soap dispensers, hip flasks, neckties, notebooks, aprons, mugs, actually just about anything can be purchased in shops or ordered online with a blue-and-white palmetto and crescent. In 2006, the *Greenwood Index-Journal* ran an article discussing the popularity of the graphic elements of the South Carolina flag, and it also questioned the flag variants and their uses: "But with so many representations of the state flag, the question remains, who has the right to use the emblem on merchandise? According to Terry Cowling, of the South Carolina Department of Parks, Recreation and Tourism, 'the state flag is a public entity, anyone can take a part of the flag and reproduce it since it's public domain.' Michael Kelly, of the South Carolina Secretary of State's Office, reiterated Cowling's statement: 'The state flag and seal are both public domain. It (the flag) has been public domain since it was adopted.'"[32] The article, with its authoritative quotes, ratified the proposition that anyone (presumably even a North Carolinian or Georgian!) was free to create their own South Carolina flag, so long as it adhered to the basics. The article went on to observe "there are few laws protecting the state flag and none making desecration of the flag and its emblems illegal."[33]

Then, starting in June 2015, presumably settled flag facts started becoming flag controversies. Black activist Bree Newsome climbed a flagpole outside the statehouse and removed the Confederate flag. In some press photos a nearby palmetto looks on mutely. Her dramatic action signaled a new level of concerns about flags and renewed sensitivity about slavery.

In May 2018, "Big Red" was back in the news. Union soldiers had taken the historic flag back to Iowa at the end of the Civil War. In 2010, Iowa took the flag out of storage and loaned it to the Citadel, where it was put on display. Even though the loan was scheduled to last until 2021, South Carolina's governor, Henry McMaster, requested that the loan be made permanent. Some black Iowan legislators wanted Iowa governor "Kim Reynolds to talk to the state's African-American community before making a decision." South Carolinians point out that "Big Red" was chosen to replace the controversial Confederate flag as the Citadel's "spirit flag."[34]

On April 29, 2017, *The State* newspaper ran a letter to the editor under the heading, "Could we please stop changing SC flag?"[35] The author was Scott Malyerck, and it seemed like the type of letter that might motivate a legislator to introduce legislation to address his concerns, which were the existence of different flags and his desire for consistency. Curiously, such legislation (SB 559) had been introduced six weeks earlier, on March 16, by Republican Senator Ronnie Cromer.[36] Companion legislation had been introduced in the House by Republican Representatives Martin, Pope, Delleney, and Huggins on April 26. So Malyerck's letter, rather than being a starting point, was actually part of a campaign to move legislation forward. Malyerck is frequently cited as a Newberry political consultant, but he can also be called a former executive director of the South Carolina Republican Party.[37] Whatever the small government party had in mind, Malyerck's letter and the pending legislation led to a flurry of articles (including an earlier version of this essay)[38] in a variety of South Carolina newspapers.

Although HB 4201 would not prevent people from creating their own version of the state flag or "infringe on any trademarked designs of a palmetto tree held by a private party," it would create a process leading to the director of the Department of Administration recommending a single-palmetto-tree design that, upon approval by concurrent resolution, would become the official, approved design to be used upon the flag flown over the statehouse and "for other public purposes."[39]

On January 24, 2018, *The State* newspaper ran an article featuring Mr. Malyerck, who had testified before a Senate panel. The chairman Democrat, John Scott, was quoted as saying, "the issue needed more research," but he hopes "the Legislature can resolve it before the end of its session this spring."[40]

Mr. Malyerck started a Facebook page, Pick SC Flag, which offered four choices, all South Carolina flags already in production. Joint Resolution 1002 was introduced on March 13, 2018, voted out of the Senate, and on March 1 it was sent to the House Committee on the Judiciary.[41] On March 16 the Facebook page reported, "It appears as if the leadership in the House of Representatives is holding up the 'state flag' bill."[42] But by November 2018 a State Flag Study Committee had been created (with Mr. Malyerck as a member) with a mission: "propose an official, uniform design for the state flag based on historically accurate details and legislative adoptions."[43] The flag study committee issued a report on March 4, 2020. Their palmetto recommendation: an interpretation of a quick pencil sketch done by an amateur artist (the aforementioned Ellen Heyward Jervey) from sometime around 1910.[44] Who knows when this official flag matter will be resolved, but since the proposed palmetto bears more than a passing resemblance to a bouquet of blown-out broken umbrella skeletons, I won't be surprised if it is not met with universal public acclaim.

In an era of rigorous branding supervised by identity-obsessed marketers fiercely pursuing compliance with templates and specified pantone colors (and legal threats?), it seemed both refreshing and historically appropriate that South Carolina bestows upon its citizenry (and flag manufacturers) the latitude to depict the state tree in whatever manner they choose so long as an upright white palmetto is in the center and that devilish "moon" is in the upper left.

In my opinion, some versions of the palmetto have become so distorted that they no longer closely resemble the state tree. But that's okay. The iconic white palmetto and crescent resplendent on a blue background have taken on more meaning and more import than the flag itself—two paired symbols that instantly and unequivocally say South Carolina in a way that most other state flag elements never have. From the earliest days South Carolina citizens and merchants have had the latitude to interpret the palmetto and crescent as they see fit—extending a measure of agency not seen elsewhere.

Boasting a history dating to the dawn of our country, the South Carolina flag has paradoxically transcended its glorious and casual ambiguity to become possibly the most recognizable of all state icons, far more identifiable than the tightly prescribed and obsessive offerings of other states.

Coda

When John Muir was raving about cabbage palms in 1867, he had never seen a sequoia. Two years later he was in Yosemite, where he wrote, "When we try to pick out anything by itself, we find it hitched to everything else in the Universe."[1] He was no doubt familiar with the nursery rhyme/poem about the implications of a single missing horseshoe nail,[2] but he was especially prescient when he wrote his more general formulation.

Today we are familiar with basic principles of ecosystems, positive feedback loops, and the seeming inevitability of unintended consequences (the butterfly effect). From our vantage point we can see that had Muir not contracted malaria in Florida, he may never had made it to Yosemite, never seen a sequoia, and never have started the Sierra Club. A modest cabbage palm may have been his peak botanical experience.

You've read two dozen essays about cabbage palms. Where else does curiosity about them lead? Everywhere. There are nine chapters that didn't make it into the book and, no doubt, nine times that number of stories that I haven't run down. Why is there a Palmetto Avenue in Brooklyn? What's it like participating in frenetic commercial frond harvesting in the weeks before Palm Sunday? Why did palmetto weevils ruin a paint job in Sanford, Florida? What's with the cabbage palm tombstone in the graveyard of the St. Helena Parish Church in Beaufort, South Carolina? What about all the things (epibionts) that live in and on cabbage palms? Do cabbage palms really have cells that never die? The list goes on. Why did it take 125 years for the cabbage palm to make it onto Florida's state seal? Tiger Woods, Bettie Page, Calvin Coolidge—they all have accompanying cabbage palm stories. What's a lightning-thwarting cabbage palm boondoggle? And how did I manage to avoid telling the story of the palmetto tortoise beetle, the larvae of which hide in a thatch of their own dried feces while the adults can grip fronds with a force sixty times their own weight?[3]

I picked one thing out by itself and found it attached to everything and the Universe. I recommend you try it. Don't settle for what the naturalist, or historian, or teacher offers. Go beyond what everyone knows. Pull threads, follow leads. Be a little obsessive and look for connections others may have missed. You don't have to write a book—just revel in the universal connectedness of it all.

25

Lost in Space

What the Aliens Will Learn about Us

Pyramids, statues, and time capsules have all been employed in an effort to create a permanent legacy of a culture. Until 1972, no tangible human relict had left the planet, but that changed when the United States launched Pioneer 10, "Earth's first emissary into space . . . heading generally for the red star Aldebaran," and it will take more than 2 million years to get there.[1] Attached to the Pioneer spacecraft was a golden plaque designed to convey information about where the craft came from. Designed by popular astronomer Carl Sagan and Frank Drake, it featured a controversial naked couple standing next to Pioneer (for scale) and other engravings they hoped could be interpreted by any beings competent enough to find and inspect it.[2]

The Pioneer project was followed in 1977 by Voyager I and II, which contained more ambitious attempts at interstellar communication— golden records (with cartridge, needle, and playing instructions). The disks contained coded information that, properly interpreted, included both recorded greetings, music, and images. There's a nursing mother, rush hour in India, a supermarket, and a Chinese dinner party. There's Bach and Chuck Berry, but no Bob Dylan. A sequoia covered with snow, but no scene of deforestation.[3] It's a slick, if jumbled, public relations piece for our planet. So whatever life forms eventually decode this disk may be disappointed if they decide to come visit and get the whole picture. Who knows when they might arrive or what may remain of the phenomena depicted in the 116 images? But if they hurry or get lucky they can see something depicted on the golden disk: cabbage palms.

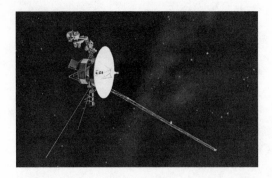

The two Voyager spacecraft carry images of cabbage palms outside our solar system. NASA image.

That's right. Of the 116 images deemed worthy of representing planet Earth, one (#113) contains two cabbage palms. It's not that Carl Sagan thought extraterrestrial beings needed to know about cabbage palms—they're simply included in the shot to frame the main subject matter: a Titan-Centaur rocket similar to the one used to launch the Voyager spacecraft. To the palms' credit there are many other pictures of the rocket that could have been chosen—but the palms form an interesting foreground and establish that, whatever else happens on this planet, there is an infinitesimal chance in a possibly infinite universe that some other culture will gaze upon image 113 and wonder why a boring, understandable rocket ship is photobombing an intriguing image of two strange life forms.

Cabbage palms frame an image of the type of Titan-Centaur rocket that launched the Voyager spacecraft. NASA image.

ACKNOWLEDGMENTS

Based on a commission I received from Mary Ruiz for a cabbage palm illustration, I've determined that I've been explicitly intrigued by these paradoxical plants since 1982. During the ensuing four decades, hundreds of
people have contributed to my fascination with this plant, and for those
not listed by name here, please accept the fact that it is my memory rather
than your lack of significance that accounts for your omission.

I owe years of incalculable gratitude to my naturalist partner and wife,
Julie Morris, and our son, Corley, for their patience, indulgence, support,
insights, navigation, and editing.

During my research and writing I relied on three men, each of whom
has made significant contribution to the understanding of cabbage palms:
Kyle Brown, Scott Zona, and David Fox.

Natural history mentors and sources of inspiration include my father,
Fin Miller, Ken Alvarez, Linda Conway Duever, Allan Horton, Robert
Dye, and Bud Kirk. Naturalist authors whose writing and general ap-
proaches to observation and storytelling have inspired me include John
McPhee, John Stilgoe, Gary Paul Nabhan, and Peter Matthiessen.

In addition to my wife, I have relied on impressive peers for their obser-
vations and support. These include Jim Beever, Jean Blackburn, Bill Burger,
Ernie Estevez, Ed Freeman, Maynard Hiss, Jean Huffman, John Lambie,
Bill Lewis, John McCarthy, Rob Patten, Belinda Perry, Cathy Salustri,
Steve Suau, and Jon Thaxton, all of whom have been reliable partners and
co-conspirators in the effort to understand and value Florida.

The book would not have been written without my education at New
College, guided by John Morrill, and my graduate work in the Florida
Studies Program of the University of South Florida, St. Petersburg, where
historian Gary Mormino helped me learn how to put the history in natural
history.

The odds are excellent that this book would not have happened without
serendipitous, catalyzing interactions involving Mary Ruiz, Susan Cerulean, and Amelia Bird.

Barbara Oehlbeck wrote the other book about cabbage palms, *The Sabal Palm: A Native Monarch,* a work I found to simultaneously be an inspiration, resource, and challenge.

Special recognition is due those individuals who spent significant time with me: John Andrews, Brian Bahder, Joy Banks, Mike Duever, Lucas England, Leslie Gaines, José Garcia, Tom Gaskins Jr., Steve Griffin, Blake Gunnels, Betty (Frazee) Haskins, Ericka Helmick, Gary Hollar, Bruce Holst, Jasmin Linkutis, Jack and Buster Longino, Peter McClure, Tim Merkel, Buddy Mills, Melissa Nell, Will Ortelt, Troy Randall, Larry Richardson, Sang Roberson, Tom Schuller, Willy the Losen, Sam and Robbie Vickers, Ron Warner, Barry Wharton, Jamie Wright, and Chappy Young.

The following additional people have made noteworthy contributions: Bob Adair, Kristy Andersen, Paula Benshoff, David Brain, Edwin Bridges, Judy Burn, Nicholas Butler, Paul Cliff, Ethel Coakley, Jeff Conkey, Cammie Donaldson, Suzanne Dorsey, Nancy Edmundson, Monica Elliot, Jason Evans, Andrew Georgiadis, Anne Gometz, Robert "Bruce" Graetz, Andrew Graham, Tom Hallock, Rusty Harley, Nigel Harrison, Mark Higgins, Al Hine, Todd Hoppock, Steve Humphrey, David Iannotti, Megan Jourdan, Rhoda Knight Kalt, Laurel Kaminsky, Gene Kelly, Stetson Kennedy, Joe Kiefer, Greg Kramer, Tom Larson, Jane Longino, Roy McBride, Caren McCourtney, Tim Merkel, Jerald Milanich, Ken Misemer, Iliana Moore, Tish Morris, James Moultrie, Joanne Mulinare, Nalini Nadkarni, Elaine Norman, Reed Noss, Jeff Palmer, Craig Pittman, Larry Rabinowitz, June Richcreek, Diane Riggleman, Wally Rockfellow, Juliet Rynear, Andy Stice, Sally Self, Tom Trotta, Howard Troxler, Jim Turner, Tom Turner, Sonny Vergara, Ron Warner, and Alice White.

In addition, the following entities deserve recognition for their support and encouragement: Florida Native Plant Society, Marie Selby Botanical Gardens, Florida Association of Native Nurseries, the former and current UF/IFAS palm pathology team, and the Florida Federation of Garden Clubs.

Finally, special prolonged appreciation is due the staff of the University Press of Florida for their decade of patience and prodding in equal measure.

NOTES

Introduction

1. John Muir, *My First Summer in the Sierra* (Boston: Houghton Mifflin, 1917), 158, http://www.gutenberg.org/files/32540/32540-h/32540-h.htm#draw7.

2. Elliot Kleinberg, "Shipwreck, its coconuts led to Palm Beach's name," Palm Beach Post, January 9, 2019, https://www.palmbeachpost.com/news/20190109/from-archives-shipwreck-its-coconuts-led-to-palm-beachs-name/.

3. Patti Anderson, "Native Palms of Florida: Identification and Regulation," Florida Department of Agriculture and Consumer Services, Division of Plant Industry, December 2019, https://www.fdacs.gov/ezs3download/download/89912/2571089/Media/Files/Plant-Industry-Files/CIRCULAR-Native-Palms-of-Florida_BH.pdf.

4. Russell M. Burns and Barbara H. Honkala, *Silvics of North America,* vol. 2: *Hardwoods* (Washington, DC: Forest Service USDA, 1990), 763, https://www.srs.fs.usda.gov/pubs/misc/ag_654/volume_2/silvics_v2.pdf.

5. John Muir, *A Thousand-Mile Walk to the Gulf* (Boston: Houghton Mifflin, 1916), 92.

6. Ibid., 118, 119.

Chapter 1. Quest for the Ages: How Long Can These Things Live?

1. Stetson Kennedy, *Palmetto Country* (Tallahassee: Florida A&M University Press, 1989. 3).

2. Radha Krueger, "Fossil Palm," *Rare, Beautiful & Fascinating* (blog), July 12, 2016, https://www.floridamuseum.ufl.edu/100years/fossil-palm/.

3. Charles P. Daghlian, "A Review of the Fossil Record of Monocotyledons," *Botanical Review* 47, no. 4 (1981): 536.

4. M. V. Parthasarathy and P. B. Tomlinson, "Anatomical Features of Metaphloem in Stems of Sabal, Cocos and Two Other Palms," *American Journal of Botany* 54, no. 9 (1967): 1143–51, https://doi.org/10.2307/2440540.

5. Kyle E. Brown, October 16, 2009, site visit and personal communication.

6. Scott Zona and Katherine Maidman, "Growth Rates of Palms in Fairchild Tropical Garden," *Palms* 45, no. 3 (2000): 154.

7. Kelly McPherson and Kimberlyn Williams, "Establishment Growth of Cabbage Palm, Sabal palmetto (Arecaceae)," *American Journal of Botany* 83, no. 12 (1996): 1567.

8. Ibid., 1569.

9. James Trager, *The People's Chronology: A Year-by-Year Record of Human Events from Prehistory to the Present* (New York: Holt, Rinehart and Winston, 1979), 499–502.

10. George "Chappy" Young, "Death of a South Florida Surveying Icon (this is not what you think)," *American Surveyor,* November 13, 2009, https://amerisurv.com/2009/11/13/death-of-a-south-florida-surveying-icon/.

11. Florida Native Plant Society, "Sabal Palmetto," 2013, https://www.fnps.org/plant/sabal-palmetto.

12. "Sabal Palmetto (Cabbage Palm)," *Smithsonian Marine Station at Fort Pierce* (blog), https://naturalhistory2.si.edu/smsfp/irlspec/Sabal_palmet.htm.

13. B. T. Longino, field trip with ranch owner, Slidell, Florida, December 5, 2009.

14. Trager, *The People's Chronology,* 446–48.

15. Henry Washington, "Survey Notes of Henry Washington East Boundary Section 24, Township 37 S, Range 22 E. 0780203.TIF," LABINS (Land Boundary Information System), 1843, https://ftp.labins.org/glo_all/Volume78/Folder%2010%20pg%20187%20to%20218/0780203.TIF.

16. Office of the Surveyor General, "General Instructions to Deputy Surveyors," November 3, 1842, https://ftp.labins.org/GLO/glo_instructions_pdfs/GLOGTE 1842.pdf.

17. Barry Wharton, "1843 Witness Tree?," email to the author, December 7, 2009.

18. Young, George "Chappy," "RE: Longino Cabbage Palm," December 7, 2009, email to the author.

19. Barry Wharton, "RE: Cabbage Palm Survey Witness Tree," December 7, 2009, email to the author.

20. Henry Nehrling, *The Plant World in Florida* (New York: Macmillan, 1933), 160.

21. "On the Border, Part I," *Florida Memory Blog* (blog), http://www.florida memory.com/items/show/295247.

22. Daniel L. Schafer, *Anna Madgigine Jai Kingsley: African Princess, Florida Slave, Plantation Slaveowner* (Gainesville: University Press of Florida, 2003), 52.

23. National Park Service, "African Passages Museum Exhibit—Fort Sumter and Fort Moultrie National Historical Park (U.S. National Park Service)," Fort Sumter and Fort Moultrie, https://www.nps.gov/fosu/learn/news/african-passages-museum-exhibit.htm.

24. David Lee Russell, *Victory on Sullivan's Island: The British Cape Fear/Charles Town Expedition of 1776* (Haverford, PA: Infinity Publishing.com, 2002), 123, 142.

25. Russell, *Victory on Sullivan's Island,* 87.

26. Virginia S. Wood, *Live Oaking: Southern Timber for Tall Ships* (Annapolis, MD: Naval Institute Press, 1995), 14.

27. Russell, *Victory on Sullivan's Island,* 87.

28. Jim Stokely, *Fort Moultrie: Constant Defender* (Washington, DC: Division of Publications, National Park Service, U.S. Department of the Interior, 1985), 15.

29. John Muir, *A Thousand-Mile Walk to the Gulf* (Boston: Houghton Mifflin, 1916), 92.

30. Edwin C. Bearss, *The Battle of Sullivan's Island and the Capture of Fort Moultrie: A Documented Narrative and Troop Movement Maps, Fort Sumter National Monument, South Carolina* (Washington, DC: National Park Service, 1969), 96.

31. Stokeley, *Fort Moultrie,* 26.

32. "Misheard Lyrics, Performed by Star Spangled Banner," amIright, http://www.amiright.com/misheard/song/starspangledbanner.shtml.

33. "Southern Theater 1780–1783," American Battlefield Trust, January 27, 2017, https://www.battlefields.org/learn/revolutionary-war/southern-theater-1780-1783.

Chapter 2. Are Cabbage Palms Trees? Wood? Grass?

1. "FindLaw's United States Supreme Court Case and Opinions," Findlaw, https://caselaw.findlaw.com/us-supreme-court/378/184.html.

2. Harold David Harrington and L. W Durrell, *How to Identify Plants* (Athens: Swallow Press of Ohio University Press, 1996).

3. Gil Nelson, *The Trees of Florida: A Reference and Field Guide* (Sarasota, FL: Pineapple, 1994), 86.

4. Penny Carnathan, "Gardener, 97, Patents Irrigation System Plants Love, Mosquitoes Don't," *Tampa Bay Times,* February 7, 2013, https://www.tampabay.com/features/homeandgarden/gardener-97-patents-irrigation-system-plants-love-mosquitoes-dont/1273900.

5. "Poaceae—Distribution and Abundance | Britannica.Com," https://www.britannica.com/plant/Poaceae/Distribution-and-abundance.

6. "Arecaceae—The Plant List," http://www.theplantlist.org/browse/A/Arecaceae/.

7. Dave Howard, "Cereals: Importance and Composition," *Thought for Food Blog,* June 21, 2012, https://www.ifis.org/blog/2012/food-science-and-technology/cereals-importance-and-composition.

8. "Life Forms," Radboud University Nijmegen, https://www.vcbio.science.ru.nl/en/virtuallessons/landscape/raunkiaer/.

9. Robert W. Long and Olga Lakela, *A Flora of Tropical Florida: A Manual of the Seed Plants and Ferns of Southern Peninsular Florida* (Coral Gables, FL: University of Miami Press, 1971), 243.

10. F. Andre Michaux, *North American Sylva* [Michaux], vol. 3: *Natural History–Original Investigations,* 5, http://lhldigital.lindahall.org/cdm/ref/collection/nat_hist/id/15461.

11. John Kunkel Small, "The Cabbage Tree: Sabal Palmetto," *Journal of the New York Botanical Garden* 24 (1923): 145–58.

12. William Bartram and Francis Harper, *The Travels of William Bartram,* Naturalist's ed. (Athens: University of Georgia Press, 1998), 89.

13. Henry Nehrling, *The Plant World in Florida* (New York: Macmillan, 1933), 160.

14. John C. Gifford, "Some Reflections on the South Florida of Long Ago," *Tequesta* 1 (1946): 6.

15. Richard Moyroud, "Cabbage Palms," *Palmetto* 16, no. 3 (1996).

16. Elbert L. Little Jr., *Atlas of United States Trees,* vol. 5: *Florida* (Washington DC: U.S. Government Printing Office, 1978), 2, 133.

17. Frank C. Craighead, *The Trees of South Florida* (Coral Gables, FL: University of Miami Press, 1971), 97.

18. David Allen Sibley, *The Sibley Guide to Trees* (New York: Knopf, 2009), 82.

19. Daniel B. Ward and Robert T. Ing, *Big Trees: The Florida Register* (Orlando: Florida Native Plant Society, 1997), 83, 142.

20. Nelson, *The Trees of Florida,* 130.

21. Team Selby, "Cabbage Palm, Sabal Palm (Palm Family)," *Marie Selby Botanical Gardens* (blog), June 30, 2015, https://selby.org/cabbage-palm-sabal-palm-palm-family/.

22. Long and Lakela, *A Flora,* 30.

Chapter 3. Bootjacks: Fronds, How They Work (and Fail)

1. Scott Zona, "A Monograph of Sabal," 587, https://doi.org/10.5642/aliso.19901204.02.

2. John Muir, *A Thousand-Mile Walk to the Gulf* (Boston: Houghton Mifflin, 1916), 118.

3. Harriet Beecher Stowe, *Palmetto Leaves* (Gainesville: University Press of Florida, 1999), 209.

4. Ibid., 253.

5. Timothy K. Broschat, Alan W. Meerow, and Monica L Elliott, *Ornamental Palm Horticulture,* 2nd ed. (Gainesville: University Press of Florida, 2017), 86, http://www.plantapalm.com/vpe/horticulture/palm-nutrition.htm

6. Dale D. Wade and O. Gordon Langdon, "Sabal Palmetto," https://www.srs.fs.usda.gov/pubs/misc/ag_654/volume_2/sabal/palmetto.htm.

7. John Kunkel Small, "The Cabbage Tree: Sabal Palmetto," *Journal of the New York Botanical Garden* 24 (1923): 158.

8. Timothy K. Broschat and Monica L. Elliot, "Normal 'Abnormalities' in Palms," UF/IFAS Extension Service, December 2017, https://edis.ifas.ufl.edu/ep344.

9. Scott Zona, "A Monograph of Sabal," 587.

10. Broschat, Meerow, and Elliott, *Ornamental Palm Horticulture,* 86.

11. John Longino, "Bull Ants and Movies," June 22, 2015, email to the author.

12. "Pileated Woodpecker Life History," All about Birds, Cornell Lab of Ornithology, https://www.allaboutbirds.org/guide/Pileated_Woodpecker/lifehistory.

13. "Florida Carpenter Ant—Camponotus Floridanus (Buckley) and Camponotus Tortuganus (Emery)," http://entnemdept.ufl.edu/creatures/urban/ants/fl_carpenter_ants.htm.

14. Iliana Moore, "Observations of Epibiont Diversity on Sabal Palmetto and Investigation of Epibionts as Suspects for Palmetto's Recent Leaf Death and Trunk Desiccation," independent study project, January 2016, New College of Florida, Sarasota.

15. Laurel Kaminsky, "DNA Sequence from That Fungus," personal communication with the author, March 22, 2018.

16. Moore, "Observations of Epibiont Diversity."

17. David A. Fox, "Sabal Palmetto: Investigating the Ecological Importance of Florida's State Tree" (Ph.D. diss., University of Florida, 2015), 79.

18. Fox, "Sabal Palmetto," 31.

19. Reed F. Noss, *Fire Ecology of Florida and the Southeastern Coastal Plain* (Gainesville: University Press of Florida, 2018), 238.

Chapter 4. Hunting Island: The Curious Recurved Seedling

1. Kelly McPherson and Kimberlyn Williams, "Establishment Growth of Cabbage Palm, Sabal Palmetto," *American Journal of Botany* 83, no. 12 (1996): 1566.

2. Kimberlyn Williams, Katherine C. Ewel, Richard P. Stumpf, Francis E. Putz, and Thomas W. Workman, "Sea-Level Rise and Coastal Forest Retreat on the West Coast of Florida, USA," *Ecology* 80, no. 6 (1999): 2045–63.

3. McPherson and Williams, "Establishment Growth," 1566.

4. P. B. Tomlinson, *The Structural Biology of Palms* (Oxford, England : Clarendon , 1990), 83.

5. John Muir, *A Thousand-Mile Walk to the Gulf* (Boston: Houghton Mifflin, 1916), 117.

Chapter 5. One Plant, Many Names

1. Carl von Linné, Matthew Baillie, James Edward Smith, and Royal College of Physicians of London, *A Selection of the Correspondence of Linnaeus and Other Naturalists,* vol. 1 (London : Longman, Hurst, Rees, Orme, and Brown, 1821), 182, https://www.biodiversitylibrary.org/item/239601.

2. Timothy Van Deelen, "Serenoa Repens," US Forest Service, Fire Effects Information System (FEIS), 1991, https://www.fs.fed.us/database/feis/plants/shrub/serrep/all.html.

3. "Historic Haile Homestead," hailehomestead, https://www.hailehomestead.org.

4. "History Kanapaha Botanical Gardens," Kanapaha Botanical Gardens, 2018, https://web.archive.org/web/20190709205940/http://kanapaha.org/home/history/.

5. Angela T. Leiva Sánchez, *Las palmas en Cuba* (Havana, Cuba: Editorial Científico-Técnica, 1999).

6. Daniel F. Austin and P. Narodny Honychurch, *Florida Ethnobotany: Fairchild Tropical Garden, Coral Gables, Florida Arizona-Sonora Desert Museum, Tucson, Arizona: With More Than 500 Species Illustrated by Penelope N. Honychurch . . . [et al.]* (Boca Raton, FL: CRC, 2004), 585, 586.

7. Debra Kay Harper, "English/Seminole Vocabulary," Seminole Wars Foundation, 2010, http://www.yallaha.com/documents/Seminole%20Dictionary.pdf. 20.

8. Michael Gannon, ed., *The New History of Florida* (Gainesville: University Press of Florida, 1996), 17, 21.

9. *The Journey of Alvar Cabeza de Vaca*, trans. Fanny Bandelier (New York: Allerton, 1922), 19, https://ia800504.us.archive.org/15/items/journeyofalvarnuoon/journeyofalvarnuoon.pdf.

10. Alvar Cabeza de Vaca, *The Journey*, 26.

11. Gannon, ed., *New History of Florida*, 26.

12. Lawrence A. Clayton, Vernon James Knight, and Edward C. Moore, eds., *The DeSoto Chronicles:* vol. 1, 2 vols. (Tuscaloosa: University of Alabama Press, 1993), 57.

13. Gannon, ed., *New History of Florida*. 27.

14. Jerald T. Milanich, "Original Inhabitants," in *The New History of Florida*, ed. Michael Gannon(Gainesville: University Press of Florida, 1996), 9.

15. Ibid., 65.

16. Alan W. Meerow, *Betrock's Cold Hardy Palms* (Hollywood, FL: Betrock Information Systems, 2005), 42, 43.

17. Jody Haynes and John McLaughlin, "Edible Palms and Their Uses Fact Sheet MDCE-00-50," UF/Miami-Dade County Extension, November 2000, http://www.quisqualis.com/tvo1ediblepalms.html.

18. John Lawson, "John Lawson, 1674–1711: A New Voyage to Carolina; Documenting the American South," 7, https://docsouth.unc.edu/nc/lawson/lawson.html.

19. Mark Catesby, *The Natural History of Carolina, Florida and the Bahama Islands*, vol. 1 (Printed for B. White, 1731), xli.

20. Mark Catesby, *Hortus Europae Americanus* (J. Millan, 1767), 40, https://archive.org/details/mobot31753000795341/.

21. Scott Zona, "A Monograph of Sabal," 621, https://doi.org/10.5642/aliso.19901204.02.

22. John Bartram, "December 31, 1765—The Journal of John Bartram," Florida History Online, https://www.unf.edu/floridahistoryonline/Bartram/December_1765/31dec1765.htm.

23. Zona, "A Monograph of Sabal," 622.

24. Scott Zona, Re: Thanks so much! Personal Communication (email) November 24. 2019.

25. David A. Fox, "Sabal Palmetto: Investigating the Ecological Importance of Florida's State Tree" (Ph.D. diss., University of Florida, 2015), 12.

26. Ibid.

27. Jack Scheper, "Sabal Palmetto," Floridata, https://floridata.com/plant/100.

28. Green Deane, "Cabbage Palm, Sabal Palmetto," Eat the Weeds and Other Things, Too, http://www.eattheweeds.com/search/sabal/.

29. David George Haskell, *The Songs of Trees: Stories from Nature's Great Connectors* (New York: Penguin, 2018), 65.

30. Leiva Sánchez, *Las palmas en Cuba,* 58.

Chapter 6. Anomalies: Oddball Palms Get Noticed (and Collected)

1. Joyce Black, "Carriers' Garden Home One of 5 on Garden Tour," *Fort Myers News-Press,* March 14, 1968.

2. Christopher O'Donnell, "A Palm Tree Worth 20 Grand?," *Sarasota Herald-Tribune,* September 22, 2007, https://www.heraldtribune.com/article/20070922/News/605233128.

3. Robert Riefer and Scott Zona, "A New Cultivar of Sabal Palmetto," *Palms* 49, no. 1 (2005): 46, 47.

4. John Kunkel Small, *Palm with Split Trunk,* January 1927, Florida Memory, State Library and Archives of Florida, https://www.floridamemory.com/items/show/50479.

5. John Kunkel Small, *Palm with Split (Two Trunks),* 19-, Florida Memory, State Library and Archives of Florida, https://www.floridamemory.com/items/show/50939.

6. "Two-Headed Palm," *Fort Myers News-Press,* November 19, 1965, https://www.newspapers.com/clip/10450132/newspress/.

7. "Exhibit A. Future Land Use Strategies Platted Lands Problems and Solutions (City of Cape Coral Evaluation and Appraisal Report)," City of Cape Coral, Florida, 2005, http://www.capecoral.net/department/community_development/comprehensive_planning/docs/exhibit_a.pdf.

8. "Cape Coral Garden Offers Big Variety of Attractions," *Fort Myers News-Press,* February 18, 1969, 110, https://www.newspapers.com/clip/9803177/twoheaded_palm_cape_coral_gardens/.

9. Bill Hutchinson, "Jelks Devotes Time, Money, Energy to the Myakka," *Sarasota Herald-Tribune,* July 17, 2005, https://www.heraldtribune.com/article/20050717/News/605249513.

10. Joan Berry, "Searching for Indian Language Trees: First-Person Account," 2018, https://www.researchgate.net/publication/323401344_SEARCHING_FOR_INDIAN_LANGUAGE_TREES_FIRST-PERSON_ACCOUNT.

11. Jerald T. Milanich and Nina J. Root, *Enchantments: Julian Dimock's Photographs of Southwest Florida* (Gainesville: University Press of Florida, 2013), 30.

12. "Two Koreshan Women in Front of Split Palm," n.d., DigitalFGCU, Florida Gulf Coast University's Digital Repository, http://purl.flvc.org/fcla/ic/SW00002294.

13. Peppersass, "Alafia River Corridor Nature Preserve, Florida," *Summit Hiking in New England* (blog), February 25, 2017.

14. "College Arms Hotel—DeLand, Florida," postcard, ca. 1910, Florida Memory, State Library and Archives of Florida, https://www.floridamemory.com/items/show/139578.

15. Bill Dipaolo, "Gardens Celebrates Founder, and 50 Years of Growing Up," *Palm Beach Post,* March 31, 2012, B002, https://www.newspapers.com/clip/33599416/former_two_headed_palm/.

16. "Two Headed Sabal Palmetto Palm Bay Fl.," *Palmpedia* (blog), n.d., http://www.palmpedia.net/wiki/Sabal_palmetto.

17. "Rare Palm Donated to Community," NBC-2, June 27, 2012.

18. Keri Byrum, "Cypress Gardens," *Miss Smarty Plants* (blog), May 17, 2015, http://misssmartyplants.com/cypress-gardens/.

19. Judy Hardiman, "Ravine Gardens State Park," *Hardiman Images* (blog), May 10, 2015, https://www.hardimanimages.com/ravine-gardens-state-park/.

20. Chris Lorry, "Some Pics from McKee Botanical Gardens," *Dave's Garden,* April 13, 2007, https://davesgarden.com/community/forums/t/712296/#b.

21. M. J. Brown and J. Nowak, "Forests of Florida, 2015 Resource Update FS-13," United States Department of Agriculture U.S. Department of Agriculture Forest Service, Southern Research Stat, 2017, https://www.fs.usda.gov/srsfia/states/fl/RU-FS-137(FL).pdf.

22. John Kunkel Small, *Palmetto with Three Branches,* 19-, Florida Memory, State Library and Archives of Florida, https://www.floridamemory.com/items/show/50675.

23. Eric Schmidt, *Sabal Palmetto,* color photograph, *Palmpedia,* 2012, http://www.palmpedia.net/wiki/File:100_6981.jpg.

24. "4-Headed Sabal Palm at Springs," *Orlando Sentinel,* June 12, 1969, 70.

25. Mike Mahan, "Rare Four-Crowned Sabal Palm Sprouts Offers for Couples," *Tampa Tribune,* May 9, 1987, 45.

26. John Kunkel Small, *Sabal Palmetto,* 1917, Florida Memory, State Library and Archives of Florida, https://www.floridamemory.com/items/show/49387.

27. Sandy Traub, "Why Tybee Island GA?," *The Lighthouse Inn* (blog), 2018, http://www.tybeebb.com/discover-tybee-island-ga/why-tybee-island-ga/.

28. "Fripp Island Golf & Beach Resort," Fripp Island Golf & Beach Resort, September 17, 2018, https://www.facebook.com/FrippIslandResort/photos/a.15772004 1782/10155512653286783/?type=1&theater.

29. Milt Salamon, "Odd Palm Tree Baffles Landscaper," *Florida Today,* July 12, 1994, 10.

30. "Creating a Two-Headed Palm Tree," PalmCo, n.d., http://www.palmco.com/double-palm-tree.html.

31. Timothy K. Broschat, Alan W. Meerow, and Monica L. Elliott, *Ornamental Palm Horticulture,* 2nd ed. (Gainesville: University Press of Florida, 2017), 185.

32. Ibid.

33. Verne Huser, *River Running* (Seattle: Mountaineers Books, 2001), 102.

34. John Kunkel Small, *Sabal Palmetto Burned by Forest Fire,* 1917, Florida Memory, State Library and Archives of Florida Memory, https://www.floridamemory.com/items/show/49464.

35. David A. Fox, "Sabal Palmetto: Investigating the Ecological Importance of Florida's State Tree" (Ph.D. diss., University of Florida, 2015), 83.

36. Timothy K. Broschat and Monica L. Elliot, "Normal 'Abnormalities' in Palms," UF/IFAS Extension Service, December 2017, https://edis.ifas.ufl.edu/ep344.

Chapter 7. Diseases Stalk the Cabbage Palm

1. Cora Mollen and Larry Weber, *Fascinating Fungi of the North Woods,* 2nd ed. (Duluth: Kollath+Stensaas, 2012), 5.

2. Monica L. Elliott, "Thielaviopsis Trunk Rot of Palm," University of Florida IFAS Extension, January 2018, https://edis.ifas.ufl.edu/pp143.

3. Ibid.

4. Monica L. Elliott and Timothy K. Broschat, "Ganoderma Butt Rot of Palms," University of Florida IFAS Extension, April 4, 2018, https://edis.ifas.ufl.edu/pp100.

5. Ibid.

6. Timothy K. Broschat, Alan W. Meerow, and Monica L Elliott, *Ornamental Palm Horticulture,* 2nd ed. (Gainesville: University Press of Florida, 2017), 144.

7. Brian W. Bahder and Ericka E. Helmick, "Lethal Yellowing (LY) of Palm," University of Florida IFAS Extension, December 17, 2018, https://edis.ifas.ufl.edu/pp146.

8. Nigel A. Harrison and Monica L. Elliott, "Texas Phoenix Palm Decline," *PP243,* 2017, 6.

9. Ibid.

10. Ibid.

11. S. E. Brown, B. O. Been, and W. A. McLaughlin, "First Report of the Presence of the Lethal Yellowing Group (16Sr IV) of Phytoplasmas in the Weeds Emilia Fosbergii and Synedrella Nodiflora in Jamaica," *Plant Pathology* 57, no. 4 (2008): 770–70, https://bsppjournals.onlinelibrary.wiley.com/doi/full/10.1111/j.1365-3059.2007.01792.x.

12. Associated Press, "Unknown Disease Killing off Florida's State Tree," Environment on NBC News, July 23, 2008, http://www.nbcnews.com/id/25818182/ns/us_news-environment/t/unknown-disease-killing-floridas-state-tree/.

13. Adam Walser, "Palms Planted under FDOT's Multi-Million Dollar Landscaping Plan Dying from Disease," ABC Action News WFTS Tampa Bay, July 10, 2017, https://www.abcactionnews.com/news/local-news/i-team-investigates/palms-planted-under-florida-dots-multi-million-dollar-landscaping-plan-dying-from-disease.

Chapter 8. Groves: The Strange Relationship with Citrus

1. Helen Harcourt, *Florida Fruits: How to Raise Them* (Louisville, KY: John P. Morton, 1886), 56.

2. Tom Schuller, phone interview by the author, October 2010.

3. Larry Keith Jackson and Louis W. Ziegler, *Citrus Growing in Florida,* 3rd ed. (Gainesville: University of Florida Press, 1991), 105.

4. "USDA APHIS | Citrus Greening," accessed May 24, 2019, https://www.aphis.usda.gov/aphis/resources/pests-diseases/hungry-pests/the-threat/citrus-greening/citrus-greening-hp.

5. Such a hammock with an "echo" of cabbage palms replacing citrus in rows may be seen using Google Earth at 28° 47′ 17.28″N and 80° 52′ 05.32″W. The Mullet Hill Farm grove lies to the east across the road.

6. Tom Schuller, personal communication with the author, October 2010; Andrew Graham, telephone conversation with the author, August 2011.

Chapter 9. Cabbage Palms in the Landscape

1. Marissa Rosenthal, ed., "Betrock's Plant Finder," Betrock Information Systems Inc., June 2019, www.PlantSearch.com, 230–234.

2. "Baldwin Park Residential Guidelines," Baldwin Park Development Company, September 2007, https://www.orlando.gov/Our-Government/Records-and-Documents/Plans-Studies/Baldwin-Park.

3. "Root Enhanced Palms," Griffin Trees, 2019, http://www.griffintrees.com/nursery/image-gallery.dot.

4. Rosenthal, ed., "Betrock's Plant Finder."

5. "Root Enhanced Palms," Griffin Trees, 2019.

6. Timothy K. Broschat, Alan W. Meerow, and Monica L. Elliott, *Ornamental Palm Horticulture,* 2nd ed. (Gainesville: University Press of Florida, 2017), 200, 201.

7. Cynthia S. Marks and George E. Marks, *Bats of Florida* (Gainesville: University Press of Florida, 2006), 69.

8. "Tree Worker Killed after Branch Snaps, Hits Him in Ormond Beach," https://www.mynews13.com/fl/orlando/news/2015/11/13/man_killed_after_tre.

9. "Two Men in Critical Condition after Ladder Falls onto Power Wires," Dripline.net, July 13, 2016, http://dripline.net/two-men-in-critical-condition-after-ladder-falls-onto-power-wires/.

10. "Local Business Owner Dies after Falling from a Palm Tree," Dripline.net, December 28, 2016, http://dripline.net/local-business-owner-dies-falling-palm-tree/.

11. "Report: Wasps Swarmed Tree Trimmer before Fatal Fall," *US News & World Report,* https://www.usnews.com/news/best-states/florida/articles/2018-08-07/report-wasps-swarmed-tree-trimmer-before-fatal-fall.

12. Liz Hardaway, "Teen Amputee Reflects on Accident Last Year, Readies for Fishing Tournament," Sun Newspapers, July 26, 2019, https://www.yoursun.com/charlotte/teen-amputee-reflects-on-accident-last-year-readies-for-fishing/article_a5d4356e-ae16-11e9-b733-130b8cf086e4.html.

13. Michael B. Sauter and Charles Stockdale, "25 Most Dangerous Jobs in America—24/7 Wall St.," January 2, 2019, https://247wallst.com/special-report/2019/01/02/25-most-dangerous-jobs-in-america-2/.

14. Bob Polomski and Debbie Shaughnessy, "Pruning Trees Factsheet HGIC 1003," Home & Garden Information Center | Clemson University, South Carolina, January 19, 1999, https://hgic.clemson.edu/factsheet/pruning-trees/.

15. Paul Craft and Denny Schrock, *All about Palms* (Des Moines, IA: Meredith Books, 2008), 36.

16. Timothy K. Broschat, "Pruning Palms," January 30, 2018, https://edis.ifas.ufl.edu/ep443.

17. Andrew Henderson, Evolution and Ecology of Palms (New York: New York Botanical Garden Press, 200), 66.

18. Craft and Schrock, *All about Palms,* 36.

19. Pamela Crawford, *Stormscaping: Landscaping to Minimize Wind Damage in Florida,* Florida Gardening Series, vol. 3 (Lake Worth, FL: Color Garden, 2005), 42, 82.

20. Broschat, Meerow, and Elliott, *Ornamental Palm Horticulture,* 201.

21. Ibid., 200.

22. "Trimmed Palm Trees Causing Uproar in St. Augustine," WTLV, accessed May 27, 2019, https://www.firstcoastnews.com/article/news/trimmed-palm-trees-causing-uproar-in-st-augustine/77-556824315.

23. Timothy K. Broschat and Edward F. Gilman, "Effects of Fertilization and Pruning on Leaf Canopy Number and Potassium Deficiency Symptom Severity in Sabal Palmetto," *Palms* 57, no. 2 (2013): 5.

24. Pruning Palms, http://edis.ifas.ufl.edu/ep443.

25. Nancy Doubrava and Bob Polomski (revisions by Carlin Munnerlyn and Joey Williamson), "Palms & Cycads," Home & Garden Information Center Factsheet 1019 | Updated, December 1, 2007, Clemson University, South Carolina, https://hgic.clemson.edu/factsheet/palms-cycads/.

Chapter 10. Picayune: Natural Florida Threatened by Native Palms

1. Bill Belleville, "Lost City of the Everglades Golden Gate Estates Was Going to Be the City of the Future: Too Bad the Land-Scam Artists Didn't Make Good on Their Promise," Sun-Sentinel.com, February 2, 1992, https://www.sun-sentinel.com/news/fl-xpm-1992-02-02-9201060511-story.html.

2. EvergladesplanOrg, *Transforming the Picayune Strand,* 2011, https://www.youtube.com/watch?v=UE8uiNIP4qU.

3. Belleville, "Lost City of the Everglades Golden Gate Estates."

4. Mike Duever, "RE: Southern Golden Gate Estates," February 13, 2020.

5. David A. Fox, "Sabal Palmetto: Investigating the Ecological Importance of Florida's State Tree" (Ph.D. diss., University of Florida, 2015), 13.

6. Ibid., 43, 44.

7. Reed F. Noss, "(5) Florida Flora and Ecosystematics," Facebook Group, Florida Flora and Ecosytematics, January 29, 2018.

8. Ginny Stibolt, "Invasive vs. Aggressive: They Are Not the Same," *Invasive vs. Aggressive* (blog), September 20, 2013, http://fnpsblog.blogspot.com/2013/09/invasive-vs-aggressive-they-are-not-same.html.

9. Ibid.

10. Juliet Rynear, "FNPS Annual Conference," October 9, 2018, email to the author.

Chapter 11. Panther-Killing Palms: Disturbance Alters the Food Chain

1. "Cabbage Palms Threaten Panthers at Florida Panther National Wildlife Refuge: Stimulus Funds Will Be Used to Eliminate Them—Cape-Coral-Daily-Breeze.Com," October 28, 2009, https://www.capecoralbreeze.com/news/local-news/2009/10/28/cabbage-palms-threaten-panthers-at-florida-panther-national-wildlife-refuge-stimulus-funds-will-be-used-to-eliminate-them/.

2. "Determining the Size of the Florida Panther Population," Fish and Wildlife Conservation Commission, August 2014, https://www.fws.gov/verobeach/Florida PantherRIT/20150819%20Statement%20on%20Estimating%20Panther%20Population%20Size%20-%20revised.pdf.

3. Florida Panther Recovery Team, "Florida Panther Recovery Plan, Third Revision," U.S Fish and Wildlife Service, January 31, 2006, https://web.archive.org/web/20071025124654/http://www.fws.gov/verobeach/Florida%20panther%20files/Panther%20Recovery%20Plan%202006_01_31%20-%20no%20figures.pdf.

Chapter 12. Birds and Bees: Wildlife in the Palms

1. All scrub jay facts are attributable to Jon Thaxton and were garnered over a long period.

2. Ronald L. Myers and John J. Ewel, eds., *Ecosystems of Florida* (Orlando: University of Central Florida Press, 1991), 104.

3. Scott Zona, "A Monograph of Sabal (Arecaceae: Coryphoideae)," *Aliso: A Journal of Systematic and Evolutionary Botany* 12, no. 4 (January 1, 1990): 583–666, https://scholarship.claremont.edu/aliso/vol12/iss4/2/.

4. Herbert W. Kale, ed., *Birds,* vol. 2 of *Rare and Endangered Biota of Florida* (Gainesville: University Presses of Florida, 1978), 34.

5. Personal observation by the author.

6. Personal observation by the author.

7. Barbara Oehlbeck, *The Sabal Palm: A Native Monarch* (Naples, FL: Gulfshore Communications, 1996), 11, 12.

8. Personal observation by the author.

9. Kyle E. Brown, "Ecological Life History and Geographical Distribution of the Cabbage Palm, Sabal Palmetto" (Ph.D. diss., North Carolina State University, 1973), 5.

10. Ibid., 36.

11. Ibid., 9.

12. L. K. Smith, "Cabbage Palmetto Yields Honey That Sours Quickly: Saw Palmetto Honey Excellent," *Gleanings in Bee Culture* 37 (1909): 39.

13. Rene Pratt, Harold P. Curtis Honey Co. Labelle, FL, personal communication with the author, 2010.

14. Usher Gay, "Palmetto Honey," *Savannah Bee Company* (blog), July 7, 2015.

15. "Sabal Palmetto Honey Aka Cabbage Palmetto Honey," *Heal Yourself Eat Honey!* (blog), June 2019, https://healthywithhoney.com/florida-raw-organic-honey -made-from-saw-palmetto-and-sabal-palmetto/.

16. "Sweetener Values Including Calories and Glycemic Index," *Sugar and Sweetener Guide* (blog), n.d., http://www.sugar-and-sweetener-guide.com/sweetener-values .html.

17. Lillian E. Arnold, "Some Honey Plants of Florida Bulletin 548," University of Florida, September 1954, 13, https://ufdc.ufl.edu/UF00026442/00001/2j. 2.

18. Mareese Caraballo, "Kitchen Corner," *Tampa Tribune,* September 15, 1946, C10.

19. Cynthia S. Marks and George E. Marks, *Bats of Florida* (Gainesville: University Press of Florida, 2006), 69.

20. Susan C. Morse, "Florida Panther: A Guide to Recognizing the Florida Panther, Its Tracks and Signs," Defenders of Wildlife, St. Petersburg, 2018, https:// defenders.org/sites/default/files/publications/florida_panther_identification_guide. pdf.

21. Shane D. Tedder, John J. Cox, Philip H. Crowley, and David S. Maehr, "Black Bears, Palms, and Giant Palm Weevils: An Intraguild Mutualism," *Open Ecology Journal* 5 (2012): 18–24.

22. Stephen Humphrey, director of the School of Natural Resources and Environment, University of Florida, personal communication with the author, 2010.

Chapter 13. Coastal Creep: Natural Distribution, East Coast

1. Scott Zona, "A Monograph of Sabal (Arecaceae: Coryphoideae)," *Aliso: A Journal of Systematic and Evolutionary Botany* 12, no. 4 (January 1, 1990): 583–666, 649, https://doi.org/10.5642/aliso.19901204.02.

2. Kyle E. Brown, "Ecological Studies of the Cabbage Palm, Sabal Palmetto IV, Ecology and Geographical Distribution," *Principes* 20, no. 4 (October 1976): 152–53.

3. Ibid., 153

4. F. Andre Michaux, *North American Sylva* [Michaux], vol. 3: *Natural History—Original Investigations,* 153, http://lhldigital.lindahall.org/cdm/ref/collection/ nat_hist/id/15461.

5. "A Strange American Island," *St. John Weekly News,* August 11, 1893, 2.

6. Landmark Preservation Associates, "Comprehensive Historical/Architectural Site Survey of Brunswick County, North Carolina," Brunswick County Board of Commissioners, September 2010, 1–47, http://www.brunswickcountync.gov/files/

planning/2015/04/plan_Final_Report_-_Brunswick_County_Comprehensive_Historical_Architectural_Site_Survey.pdf.

7. Take Google Earth to location 33°51'40.46"N, 77°57'36.28"W and use the Historical Imagery feature to spot cabbage palms above the maritime hammock.

8. Brown, "Ecological Studies IV," 152.

9. David A. Fox, "Sabal Palmetto: Investigating the Ecological Importance of Florida's State Tree" (PhD diss., University of Florida, 2015), 17.

10. Timothy K. Broschat, "Cold Damage on Palms," University of Florida IFAS Extension, January 31, 2018, https://edis.ifas.ufl.edu/mg318.

11. "Raleigh Breaks Record for Most Consecutive Hours of Freezing Temps," *WRAL Weathercenter Blog,* January 6, 2018, https://www.wral.com/raleigh-breaks-record-for-most-consecutive-hours-of-freezing-temps/17238393/.

12. "Possible Wild S. Palmetto in New Hanover County, NC?," PalmTalk, March 2018, https://www.palmtalk.org/forum/index.php?/topic/56202-possible-wild-s-palmetto-in-new-hanover-county-nc/.

13. David A. Francko, *Palms Won't Grow Here and Other Myths: Warm-Climate Plants for Cooler Areas* (Portland, OR: Timber, 2003), 121.

14. Ibid., 122.

Chapter 14. Coastal Creep: Natural Distribution, Gulf Coast

1. Elbert L. Little Jr., *Atlas of United States Trees,* vol. 5: *Florida* (Washington, DC: U.S. Government Printing Office, 1978).

2. Ibid., 133.

3. Kyle E. Brown, "Ecological Studies of the Cabbage Palm, Sabal Palmetto IV, Ecology and Geographical Distribution," *Principes* 20, no. 4 (October 1976): 153.

4. Brown, "Ecological Studies IV," 156.

5. Kyle E. Brown, "Ecological Studies of the Cabbage Palm, Sabal Palmetto, II. Dispersal, Predation, and Escape of Seeds," *Principes* 20, no. 2 (April 1976): 49.

6. Ibid., 51.

7. Brown, "Ecological Studies IV," 156.

8. Stephen Humphrey, "Florida Botany II (Archive) Comments by Jean Huffman, and Edwin Bridges, et al.," Facebook Groups, December 9, 2014.

9. Dale D. Wade and O. Gordon Langdon, "Sabal palmetto," in Russell M. Burns and Barbara H. Honkala, *Silvics of North America,* vol. 2: *Hardwoods,* agriculture handbook (Washington, DC: Forest Service USDA, 1990), https://www.srs.fs.usda.gov/pubs/misc/ag_654/volume_2/sabal/palmetto.htm.

10. Scott Zona, "A Monograph of Sabal," 917, https://doi.org/10.5642/aliso.19901204.02.

11. Zona, "A Monograph Of Sabal," 616.

12. Scott Zona, "RE: Kyle Brown Sabal Referral," February 1, 2014, email to the author.

13. Herman Kurz, "Florida Dunes and Scrub, Vegetation and Geology Geological Bulletin No. 23," State of Florida Department of Conservation, 1942, 141, http://palmm.digital.flvc.org/islandora/object/uf%3A64560#page/1/mode/2up.

14. "Fish, Hike, Paddle and Enjoy Birding on the Coastal Dune Lakes of South Walton | Walton Outdoors," Walton Outdoors, 2012, https://www.waltonoutdoors.com/fish-hike-paddle-and-enjoy-birding-on-the-coastal-dune-lakes-of-south-walton/.

15. Zona, "A Monograph of Sabal," 613.

Chapter 15. Last Tree Standing: Rising Sea Level on Florida's Big Bend Coast

1. Jono Miller, "The Lightning Bug and the Moth," in *The Book of the Everglades,* ed. Susan Cerulean (Minneapolis, MN: Milkweed Editions, 2002), 179.

2. Kimberlyn Williams, Katherine C. Ewel, Richard P. Stumpf, Francis E. Putz, and Thomas W. Workman, "Sea-Level Rise and Coastal Forest Retreat on the West Coast of Florida, USA," *Ecology* 80, no. 6 (1999): 2046.

3. Albert C. Hine, Daniel F. Belknap, Joan G. Hutton, Eric B. Osking, and Mark W. Evans, "Recent Geological History and Modern Sedimentary Processes along an Incipient, Low-Energy, Epicontinental-Sea Coastline; Northwest Florida," *Journal of Sedimentary Research* 58, no. 4 (July 1, 1988): 567.

4. Kimberlyn Williams, Zuleika S. Pinzon, Richard P. Stumpf, and Ellen A. Raabe, "Sea Level Rise and Coastal Forests of the Gulf of Mexico," Open File, USGS, Center for Coastal Geology, 1999, https://pubs.usgs.gov/of/1999/0441/report.pdf.

5. John Muir, *A Thousand-Mile Walk to the Gulf* (Boston: Houghton Mifflin, 1916), 92.

6. David L. Jones, *Palms throughout the World* (Washington, DC: Smithsonian Institution Press, 1995), 275–76.

7. Ronald L. Myers and John J. Ewel, eds., *Ecosystems of Florida* (Orlando: University of Central Florida Press, 1991), 519.

8. Williams, Pinzon, et al., "Sea Level Rise and Coastal Forests of the Gulf of Mexico," 40.

9. Russell M. Burns and Barbara H. Honkala, *Silvics of North America,* vol. 2: *Hardwoods,* agriculture handbook (Washington, DC: Forest Service USDA, 1990), https://www.srs.fs.usda.gov/pubs/misc/ag_654/volume_2/silvics_v2.pdf.

10. Williams, Pinzon, et al., "Sea Level Rise and Coastal Forests of the Gulf of Mexico," 7, 8.

11. Williams, Ewel, et al., "Sea Level Rise and Coastal Forest Retreat on the West Coast of Florida, USA," 2059.

12. Ibid., 2046.

13. Ibid., 2058.

14. Ibid., 1566.

15. Williams, Pinzon, et al., "Sea Level Rise and Coastal Forests of the Gulf of Mexico," 17.

16. Ibid., 34.

17. Williams, Ewel, et al., "Sea Level Rise and Coastal Forest Retreat on the West Coast of Florida, USA," 2049.

18. Williams, Pinzon, et al., "Sea Level Rise and Coastal Forests of the Gulf of Mexico, 19.

19. Williams, Ewel, et al., "Sea Level Rise and Coastal Forest Retreat on the West Coast of Florida, USA."

20. Mike Szabados, "Understanding Sea Level Change," *ASCM Bulletin* 236 (December 2008): 6.

21. Virginia R. Burkett, Douglas A. Wilcox, Robert Stottlemyer, Wylie Barrow, Dan Fagre, Jill Baron, Jeff Price, et al., "Nonlinear Dynamics in Ecosystem Response to Climatic Change: Case Studies and Policy Implications," *Ecological Complexity* 2, no. 4 (December 1, 2005): 376, https://doi.org/10.1016/j.ecocom.2005.04.010.

22. Smiriti Bhotika, Larisa Grawe DeSantis, Kimberlyn Williams, and Francis Putz, "Accelerated Decline of Sabal Palmetto in a Florida Coastal Forest," abstract from annual meeting of Ecological Society of America, 2006, Memphis.

Chapter 16. Hearts of Palm: The Tree You Can Eat

1. Bud Schulberg, *Across the Everglades* (New York: Random House, 1958), 27.

2. 2 FAO, "FAOSTAT CROPS (Dates)," Food and Agriculture Organization of the United Nations, n.d. http://www.fao.org/faostat/en/#data/QC/visualize.

3. "FAOSTAT CROPS (Coconut)," http://www.fao.org/faostat/en/#data/QC/visualize.

4. "Palm Oil—Deforestation for Everyday Products—Rainforest Rescue," https://www.rainforest-rescue.org/topics/palm-oil.

5. W. H. Hodge, "Bermuda's Palmetto," *Principes* 4, no. 3 (1960): 98.

6. Mark Catesby, "Hortus Europae Americanus," 1767, 40, https://archive.org/details/mobot31753000795341/page/39.

7. Hodge, "Bermuda's Palmetto," 93.

8. Green Deane, "Cabbage Palm, Sabal Palmetto," Eat the Weeds and Other Things, Too, August 31, 2011, http://www.eattheweeds.com/cabbage-palm-sabal-palmetto/.

9. Reid F. Tillery, *Surviving the Wilds of Florida* (Melrose, FL: Collingwood, 2005), 94.

10. Jonathan Dickinson, *God's Protecting Providence, Man's Surest Help and Defence* . . . (London, 1700), 29, https://archive.org/details/godsprotectingproodick.

11. Dickinson, *God's Protecting Providence*, 4.

12. Elaine Watson, "Surfing, Super Fruits and Social Justice: Sambazon and the Genesis of an Amazonian Super Food Empire," foodnavigator-usa.com, accessed June 25, 2019, https://www.foodnavigator-usa.com/Article/2013/01/03/Sambazon-How-to-build-an-Amazonian-superfood-empire.

13. Katherine Zeratsky, "Acai Berries: Do They Have Health Benefits?—Mayo

Clinic," Mayo Clinic Healthy Lifestyle Nutrition and Health, May 2, 2018, https://www.mayoclinic.org/healthy-lifestyle/nutrition-and-healthy-eating/expert-answers/acai/faq-20057794.

14. John Bartram, *A Description of East Florida with a Journal Kept by John Bartram . . .* , 3rd. ed. (1769), 20, https://palmm.digital.flvc.org/islandora/object/uf%3A45758#page/19+(2)/mode/2up.

15. A Recent Traveler in the Province, *Notices of East Florida with an Account of the Seminole Nation of Indians* (Charleston, SC: A. E. Miller, 1822), 31, https://ia802702.us.archive.org/6/items/noticesofeastflooosimm/noticesofeastflooosimm.pdf.

16. Jerald T. Milanich, *Florida's Indians from Ancient Times to the Present: Native Peoples, Cultures, and Places of the Southeastern United States* (Gainesville: University Press of Florida, 1998), 177.

17. Daniel F. Austin and P. Narodny Honychurch, *Florida Ethnobotany: Fairchild Tropical Garden, Coral Gables, Florida Arizona-Sonora Desert Museum, Tucson, Arizona: With More Than 500 Species Illustrated by Penelope N. Honychurch . . . [et al.]* (Boca Raton, FL: CRC, 2004), 586.

18. F. Andre Michaux, *North American Sylva* [Michaux], vol. 3: *Natural History—Original Investigations,* accessed May 27, 2019, http://lhldigital.lindahall.org/cdm/ref/collection/nat_hist/id/15461.

19. Bass Grocery and Market, "Bass 'Sells It for Less,'" *Tampa Tribune,* August 12, 1922, 2.

20. "Palm Cabbage Delights Visitors: 'King' Spends 30 Years at Work," *Fort Myers News-Press,* October 26, 1955, 64.

21. Charlyne Varkonyl, "Don't Palm off This Tasty Book," *Fort Lauderdale News,* November 7, 1982. 99.

22. David L. Jones, *Palms throughout the World* (Washington, DC: Smithsonian Institution Press, 1995), 57.

23. Stetson Kennedy, *Palmetto Country* (Tallahassee: Florida A&M University Press, 1989), 4

24. Marjorie Kinnan Rawlings and Robert Camp, *Cross Creek Cookery* (New York: Scribner, 1970), 65.

25. Rawlings and Camp, *Cross Creek Cookery,* 65.

26. Ernest Lyons, *My Florida* (South Brunswick, NJ: A. S. Barnes, 1969), 104.

27. Kennedy, *Palmetto Country,* 3.

28. Tom Gaskins, *Florida Facts and Fallacies* (Palmdale, FL: Tom Gaskins, 1978), 49.

29. Alice Feinstein, "Swamp Cabbage: Southern Florida Native with a Heart of Pure Palm," *Chicago Tribune,* July 23, 1987, https://www.chicagotribune.com/news/ct-xpm-1987-07-23-8702230728-story.html.

30. A Lady of Charleston, *The Carolina Housewife on House and Home* (Charleston, SC: W. R. Babcock, 1851), 183.

Chapter 17. Palm Entrepreneurs: Money Growing on Trees

1. Visit Natural North Florida, "Museum State Park," n.d., https://www.natural-northflorida.com/things-to-do/forest-capital-museum-state-park/.

2. Baynard Kendrick and Barry W. Walsh. *A History of Florida Forests* (Gainesville: University Press of Florida, 2007), 59, 193.

3. Laurence C. Walker, *The Southern Forest: A Chronicle* (Austin: University of Texas Press, 1991), 79.

4. M. J. Brown and J. Nowak, "Forests of Florida, 2015 Resource Update FS-137," U.S. Department of Agriculture Forest Service, Southern Research Station, September 2017, https://www.fs.usda.gov/srsfia/states/fl/RU-FS-137(FL).pdf.

5. Tom Gaskins, *Florida Facts and Fallacies* (Palmdale, FL: Tom Gaskins, 1978), 50.

6. Tom, "Topic: Sawing a Palm Tree @##$$%%," March 23, 2002, http://forestry-forum.com/board/index.php?topic=884.0.

7. *C. L. Baker and Edgar Huff Sawing a Palm Log—New Port Richey, Florida, 1946*, black-and-white photoprint, 4 × 5 in., Florida Memory, State Library and Archives of Florida, https://www.floridamemory.com/items/show/65195.

8. P. B. Tomlinson, James W. Horn, and Jack B. Fisher, *The Anatomy of Palms: Arecaceae—Palmae* (Oxford: Oxford University Press, 2011), 63.

9. Mark, "Re: Palm Tree Wood Reply #13," *The Forestry Forum*, December 5, 2006, http://forestryforum.com/board/index.php?topic=21038.0.

10. Lawrence S. Earley, *Looking for Longleaf: The Fall and Rise of an American Forest* (Chapel Hill: University of North Carolina Press, 2004), 2.

11. J. L. Chamberlain, A. L. Hammett, and P. A. Araman, "Non-Timber Forest Products in Sustainable Forest Management," Proceedings, Southern Forest Science Conference, 2001, https://www.srs.fs.usda.gov/pubs/VT_Publications/01t25.pdf.

12. Daniel F. Austin and P. Narodny Honychurch, *Florida Ethnobotany: Fairchild Tropical Garden, Coral Gables, Florida Arizona-Sonora Desert Museum, Tucson, Arizona: With More Than 500 Species Illustrated by Penelope N. Honychurch . . . [et al.]* (Boca Raton, FL: CRC, 2004), 586.

13. "Facts about the Palmetto Industry," *Tampa Bay Times*, September 7, 1901, 7.

14. John Champneys, "Agriculture," *North-Carolina Star*, October 12, 1809, 3.

15. N. H. Allen et al., "Wharf Building," *Pensacola Gazette*, August 24, 1827, 3, https://www.newspapers.com/clip/20508734/pensacola_gazette/.

16. Andrew Jackson, "Marine Railway &C." *Pensacola Gazette*, June 12, 1830, 1.

17. John W. Andrews, *Cedar Key Fiber and Brush Factory: From the Tree to Thee* (Ocala, FL: Atlantic Publishing Group, 2016), 35.

18. "Cabbage Palm Piling for Docks, Etc.," *Tampa Tribune*, September 18, 1960, 19C.

19. Austin, *Florida Ethnobotany*, 586.

20. "Many Plants Available in Fibre Industry for Florida Make Waste Land

Profitable," *Fort Myers News-Press*, October 27, 1926, 7, https://www.newspapers.com/clip/7078322/uses_of_fibre_palm_not_mentioned_on/.

21. Bob Prichard, "New Dunedin Firm Will Employ 1,500," *Tampa Bay Times*, August 2, 1951. 3, https://www.newspapers.com/clip/20518075/tampa_bay_times/.

22. "Tampa Firm to Establish Okeechobee Palm Plant," *Tampa Morning Tribune*, September 8, 1953, 26.

23. "3d Suspect in Stock Deal Hunted in St. Petersburg," *Miami News*, January 25, 1957, 20. https://www.newspapers.com/clip/20531001/the_miami_news/.

24. "Three Deny Swindle in Okeechobee Palm Firm," *Tampa Tribune*, February 9, 1957, 4, https://www.newspapers.com/clip/20531559/the_tampa_tribune/.

25. Henry Cavendish, "Battle over Secret Process Marks Rinehart Group Trial," *Miami News*, March 4, 1958, 12, https://www.newspapers.com/clip/16505783/the_miami_news/.

26. "Ice Cream Said 'Product' in Rinehart Firm," *Tampa Times*, March 15, 1958, 5, https://www.newspapers.com/clip/16507068/the_tampa_times/.

27. "Reports Differ on Cellulose," *Tallahassee Democrat*, March 19, 1958, 2, https://www.newspapers.com/clip/16509391/tallahassee_democrat/.

28. "Rinehart Sentenced in Palm Paper Deal," *Orlando Sentinel*, July 1, 1958, 9, https://www.newspapers.com/clip/20550772/the_orlando_sentinel/.

29. "Rinehart, Powers Cleared by Court," *Tallahassee Democrat*, September 3, 1960, 1, https://www.newspapers.com/clip/20550234/tallahassee_democrat/.

30. Bo Poertner, "Time to Brush up on DeBary History," *Orlando Sentinel*, July 1, 1994, B-1, https://www.newspapers.com/image/232445647/?terms=%22DeBary%2BHistory%22#.

31. "Jacksonville Devastated," *Montrose Democrat*, May 9, 1901, 2, https://www.newspapers.com/clip/24700612/the_montrose_democrat/.

32. Poertner, "Time to Brush Up," B-1.

33. Sally Price, "Remembering the Past," *Newscaster/Nature Coast News*, January 4, 2012, wscasterarchives.com/1-4-12/Newspg4.php.

34. Andrews, *Cedar Key Fiber and Brush Factory*, 47.

35. Ibid., 49.

36. Betty Berger, *Back Roads* (Bloomington, IN: Author House, 2008), 71.

37. Andrews, *Cedar Key Fiber*, 50.

38. Ibid., 59.

39. Ibid., 54.

40. Ibid., 72.

41. Ibid., 68.

42. Ibid., 72.

43. Ibid., 83

44. Ibid., 35.

45. Florida State Chamber of Commerce, *Industrial Directory of Florida*, 1935.

46. "Cabbage Palm Trees Are Fast Disappearing," *New Smyrna Daily News,* August 15, 1924, https://www.newspapers.com/clip/7086984/sargent_as_quoted_by_nehrling__part_1/.

47. D. A. Andrews, Design for a Brush, 61,358, Cedar Keys, FL, filed April 15, 1922, and issued August 15, 1922, https://patentimages.storage.googleapis.com/42/e3/b4/47921c5f761232/USD61358.pdf.

48. Andrews, Design for a Brush.

49. Frederick F. Strong, Process of Converting Fibrous Plants into Textile Fiber and Pulp, St. Petersburg, FL, filed June 10, 1910, and issued October 10, 1911, https://patentimages.storage.googleapis.com/c5/38/5a/79132a76ff39bc/US1005354.pdf.

50. Frank Dexter, Tire. 1,141,620, Vista, FL, filed March 30, 1914, and issued June 1, 1915, https://patentimages.storage.googleapis.com/5b/6e/a3/7c8e0764c2daff/US1141620.pdf.

51. Eugene L. Brasol, Flowerpot, 2,700,847, Daytona Beach, FL, filed May 3, 1951, and issued February 1, 1955, https://patents.google.com/patent/US2700847

52. Alex Schauss and Winson Fong, Palm Fiber-Based Dietary Supplements, World Intellectual Property Organization WO 2006/014979 A2, filed July 27, 2005, and issued February 9, 2006, https://patents.google.com/patent/US20130183391A1/en?q=%22Palm+Fiber-Based+Dietary+Supplements%22&oq=%22Palm+Fiber-Based+Dietary+Supplements%22.

53. Barbara Thorne, Termite Preferred Resource Compositions and Methods, US 2018/0006197 A1, Vero Beach, FL, filed March 24, 2016, and issued May 3, 2018, https://patentimages.storage.googleapis.com/4c/d5/b4/57c8c25d6168d8/US20180116197A1.pdf.

54. "Lowly Cabbage Palm a Most Useful Tree, Forester Says," *Tampa Tribune,* April 11, 1948, 10, https://www.newspapers.com/clip/16505373/the_tampa_tribune/.

55. Russell M. Burns and Barbara H. Honkala, *Silvics of North America,* vol. 2: *Hardwoods,* agriculture handbook (Washington, DC: Forest Service USDA, 1990), https://www.srs.fs.usda.gov/pubs/misc/ag_654/volume_2/silvics_v2.pdf.

56. "Florida Resources and Possibilities," *Palm Beach Post,* May 7, 1936, 4, https://www.newspapers.com/clip/7080580/waterproof_coffins/.

Chapter 18. Seminole Thatch: Chickees and Tikis

1. Carrie Dilley, *Thatched Roofs and Open Sides: The Architecture of Chickees and Their Changing Role in Seminole Society* (Gainesville: University Press of Florida, 2015), 14.

2. Ibid., 15.

3. Ibid., 30.

4. Ibid., 38, 33.

5. Theodor de Bry, "XXXI. How They Set on Fire an Enemy's Town," Florida Memory, State Library and Archives of Florida, https://www.floridamemory.com/items/show/294797.

6. "Theodor de Bry's Engravings of the Timucua—Collection Description," Florida Memory, State Library and Archives of Florida, https://www.floridamemory. com/find?keywords=%22de+Bry%27s+engravings%22.

7. Clay MacCauley, *The Seminole Indians of Florida,* Southeastern Classics in Archaeology, Anthropology, and History (Gainesville: University Press of Florida, 2000), 500.

8. Dilley, *Thatched Roofs*, 15.

9. Maria Herrera, "Tiki Huts Often Violate South Florida Zoning Rules," Sun-Sentinel.com, October 26, 2012, https://www.sun-sentinel.com/news/fl-xpm-2012-10-06-fl-tiki-huts-20121005-story.html.

10. "Tiki Torch Fuel Poses a Serious Danger to Children," *TulsaKids Magazine* (blog), June 10, 2012, https://www.tulsakids.com/tiki-torch-fuel-poses-a-serious -danger-to-children/.

Chapter 19. Sweetgrass Baskets: Gullah/Geechee South Carolinians Create Value from African Skills

1. "Museums | Mount Pleasant, SC—Official Website," http://www.tompsc. com/239/Museums.

2. Schuyler Kropf, "Imports Rile Basket Makers," *Post and Courier,* May 21, 2012, https://www.postandcourier.com/archives/imports-rile-basket-makers/ article_0fb81bff-7354-5895-bb48-317df7d3be4c.html.

3. Gwen McElveen, "The Role of African Slaves on South Carolina Rice Plantations," http://www.teachingushistory.org/lessons/TheRoleofAfricanSlavesonSouth CarolinaRicePlantations.html.

4. Patricia Jones-Jackson, *When Roots Die: Endangered Traditions on the Sea Islands,* 2nd printing (Athens: University of Georgia Press, 1989), 112.

5. James H. Tuten, *Lowcountry Time and Tide: The Fall of the South Carolina Rice Kingdom* (Columbia: University of South Carolina Press, 2010), https://www.sc.edu/ uscpress/books/2010/3926x.pdf.

6. Dale Rosengarten, *Row upon Row: Sea Grass Baskets of the South Carolina Lowcountry* (Columbia: McKissick Museum, University of South Carolina, 1986), 9.

7. Town of Mount Pleasant, "Appendix B Town of Mount Pleasant Ordinance—Sweetgrass Baskets," Town of Mount Pleasant, January 9, 2007, http:// www.tompsc.com/DocumentCenter/View/5065/Appendix-B_Mt-P-Ordinance _Sweetgrass-Baskets?bidId=.

8. Ann Margaret Michels, "Expanding into History—Sweetgrass Basket Stands," *Mount Pleasant Magazine,* November 3, 2011, http://mountpleasantmagazine. com/2011/places/expanding-into-history-sweetgrass-basket-stands/.

9. Viva Jane Cooke and Julia May Sampley, *Palmetto Braiding and Weaving* (Miami: E. A. Seemann, 1972), 19.

10. Rosengarten, *Row upon Row,* 12.

11. "Charleston Sweetgrass Baskets: Original and Hand-Made South Carolina Sweetgrass Baskets," CharlestonSweetgrass.com.

12. Rosengarten, *Row upon Row,* 43.

13. Kropf, "Imports Rile Basket Makers."

Chapter 20. Palm Abodes: A Different Approach to Log Cabins

1. Barbara Oehlbeck, *The Sabal Palm: A Native Monarch* (Naples, FL: Gulfshore Communications, 1996), 55.

2. Jerald T. Milanich, *Florida Indians and the Invasion from Europe* (Gainesville: University Press of Florida, 1998), 75.

3. "XXX. Construction of Fortified Towns among the Floridians," Florida Memory, State Library and Archives of Florida, https://www.floridamemory.com/items/show/294796.

4. "XXXIII. How They Declare War," Florida Memory, accessed July 3, 2019, State Library and Archives of Florida, https://www.floridamemory.com/items/show/294799.

5. Jerald T. Milanich, "The Devil in the Details," *Archaeology,* June 2005, https://archive.archaeology.org/0505/abstracts/florida.html.

6. "Talahloka National Register of Historic Places Registration Form," U.S. Dept of Interior, National Park Service, July 24, 1989.

7. Debra Kay Harper, "English/Seminole Vocabulary," Seminole Wars Foundation, 2010, http://www.yallaha.com/documents/Seminole%20Dictionary.pdf.

8. Frederick R. Barkley, "Can Palms Provide Low-Cost Homes?," *Tampa Bay Times,* February 13, 1949, 40.

9. Christopher H. Briand, "Cypress Knees: An Enduring Enigma," *Arnoldia* 60, no. 4 (2000), http://arnoldia.arboretum.harvard.edu/pdf/articles/2000-60-4-cypress-knees-an-enduring-enigma.pdf.

Chapter 21. Cheesecake: The Most-Photographed Cabbage Palm

1. "Betty Frazee Gets Prize," *Ocala Star-Banner,* August 17, 1958, 3, http://news.google.com/newspapers?id=w4BPAAAAIBAJ&sjid=6wQEAAAAIBAJ&pg=2800,595895&dq=betty-frazee&hl=en.

2. "Rotarians Hear Mozert Tell of Photo Technique," *Ocala Star-Banner,* December 30, 1958, 6. https://news.google.com/newspapers?id=quQTAAAAIBAJ&sjid=NgUEAAAAIBAJ&pg=2883,8298852&dq=betty-frazee+coke&hl=en.

Chapter 22. Fine Artists Contemplate the Cabbage Palm

1. "John Singer Sargent: Muddy Alligators," *Worcester Art Museum* (blog), https://www.worcesterart.org/collection/American/1917.86.html.

2. Singer, "John Singer Sargent's Palms," *JSS Virtual Gallery* (blog), 1917, https://www.jssgallery.org/Paintings/10119.html.

3. "Sargent: The Complete Works," *Palmettos & Palmettos, Florida* (blog), June 21, 2019, https://www.johnsingersargent.org/the-complete-works.html?q=palmetto.

4. Winslow Homer, *A 'Norther,' Key West, Fine Arts Museums of San Francisco* (blog), 1886, https://art.famsf.org/winslow-homer/norther-key-west-1979754.

5. Winslow Homer, *The White Rowboat, St. Johns River, Cummer Museum Permanent Collection* (blog), 1890, https://www.cummermuseum.org/visit/art/permanent-collection/white-rowboat-st-johns-river.

6. "Hermann Herzog and Florida," Edward and Deborah Pollack Fine Art, http://www.edwardanddeborahpollack.com/Herzogandflorida.html.

7. Herman[n] Ottomar Herzog, *Cabbage Palmettos in the Swamp, Florida, Current Artists on Display March 2016* (blog), June 21, 2019, https://slidex.tips/download/current-artists-on-display-march-2016.

8. "Selected Works of Charles R. Knight," American Profile, March 14, 2012, https://americanprofile.com/slideshows/charles-r-knight-selected-works/.

9. Richard Milner, *Charles R. Knight: The Artist Who Saw through Time* (New York: Abrams, 2012), 172.

10. John James Audubon, *Louisiana Heron,* Audubon website, November 25, 2014, https://www.audubon.org/birds-of-america/louisiana-heron.

11. "Lehman, George | Library Company of Philadelphia Digital Collections," 53, https://digital.librarycompany.org/islandora/object/digitool:79453.

Chapter 23. Toby Kwimper: A Cinematic Experiment in Ecological Succession

1. "Elvis Presley | Inglis & Yankeetown, Florida | Inglis, Florida | Yankeetown, Florida | Withlacoochee River | Florida Nature Coast," Withlacoochee Gulf Area Chamber of Commerce, http://inglisyankeetownchamber.com/Elvis-Presley.html.

2. Arlene Alligood, "Elvis About to Happen to Yankeetown," *St. Petersburg Times,* June 28, 1961, 12, https://www.newspapers.com/clip/21146355/tampa_bay_times/.

3. Barbara Marshall, "When Elvis Came to Town," *Palm Beach Post,* December 23, 2012. D12.

4. "Elvis Presley's Films, Best to Worst," *IMDb* (blog), July 4, 2013, https://www.imdb.com/list/ls053442193/; Alan Hanson, "Elvis History Blog," *Follow That Dream* (blog), April 2008, http://www.elvis-history-blog.com/follow-that-dream.html; Harley Payette, "'Elvis' Best Performances." *Elvis Information Network* (blog), June 2006, http://www.elvisinfonet.com/spotlight_best_movie_performances.html.

5. Tom Knotts, *See Yankeetown* (Ft. Lauderdale, FL: Associated Printing Corp., 1970), 17.

6. Alligood, "Elvis About to Happen to Yankeetown," 46.

Chapter 24. Elegance Doesn't Come Easy: The South Carolina State Flag

1. Ryan Hatch, "All 50 State Flags, Ranked," *Thrillist* (blog), August 13, 2014, https://www.thrillist.com/gear/all-50-us-state-flags-ranked.

2. John Nova Lomax, "For Flag Day: America's 10 Best State Flags," *Houston Press,* June 14, 2012, https://www.houstonpress.com/news/for-flag-day-americas-10-best -state-flags-6732688.

3. Edward B. Kaye, "Good Flag, Bad Flag, and the Great NAVA Flag Survey of 2001," *Raven* 8 (2001), https://nava.org/digital-library/raven/Raven_v08_2001_ p011-038.pdf.

4. Anne Wylma Wates, *A Flag Worthy of Your State and People: The History of the South Carolina State Flag,* 2nd rev. ed. (South Carolina Department of Archives and History, 1996), 2.

5. W. Eric Emerson, Paul Koch, Robert Dawkins, Scott M. Malyerck, and Walter Edgar, *Report of the South Carolina State Flag Study Committee,* March 4, 2020, 12, https://www.scstatehouse.gov/CommitteeInfo/SCStateFlagStudyCommittee/ SC%20SFSC%20Final%20Report-merged.pdf.

6. "Source of Crescent and Tree on the South Carolina Flag? (U.S.)," FOTW Flags of the World, accessed June 26, 2019.

7. "South Carolina State House | South Carolina State Symbols," https://www.sc statehouse.gov/studentpage/coolstuff/seal.shtml.

8. Richard W. Hatcher, "The Problem in Charleston Harbor," *Hallowed Ground Magazine,* 2010, https://www.battlefields.org/learn/articles/problem-charleston -harbor.

9. Ted "Tex" Curtis, "Big Red Background," citadelalumni, June 2009, https://issuu .com/citadelalumni/docs/big-red-background.

10. Ibid.

11. "'Big Red' Comes Home," *The Citadel Newsroom,* March 5, 2010, https://web. archive.org/web/20190903153721/http://www.citadel.edu/root/bigred_returns.

12. "The Palmetto Guard Flag," The Civil War in South Carolina, 1998, http:// www.researchonline.net/sccw/conflag6.htm; "Flags of Fort Sumter," Wayback Machine, November 8, 2018, U.S. Department of the Interior, National Park Service, https://web.archive.org/web/20181108071507/https://www.nps.gov/fosu/planyour-visit/upload/Flags_of_Fort_Sumter.pdf.

13. Glenn Dedmondt, *The Flags of Civil War South Carolina* (Gretna, LA: Pelican, 2000), 16, 17.

14. Ted Kaye, "Good Flag, Bad Flag: How to Design a Great Flag," North American Vexillological Association, n.d., https://www.ausflag.com.au/assets/images/ Good-Flag-Bad-Flag.pdf.

15. Wates, *A Flag Worthy,* 8.

16. Ibid., 10.

17. Ibid., 11.

18. Ibid., 3.

19. Ibid., 15.

20. "A Splendid Token," *Orangeburg Times and Democrat,* July 15, 1913, 1.

21. An Act to provide for the display of the state flag over public buildings, Pub. L. No. 406, § Acts and Joint Resolutions of the General Assembly of the State of South Carolina at the Regular Session of 1910 (1910), 753, https://babel.hathitrust.org/cgi/pt?id=iau.31858020992958&view=1up&seq=271.

22. Wates, *A Flag Worthy*, 18.

23. Robert Behre, "The History behind Why South Carolina Has No Formal Template for Its Official State Flag," *Charleston Post and Courier*, May 12, 2018, https://www.postandcourier.com/news/the-history-behind-why-south-carolina-has-no-formal-template/article_f46ed714-5490-11e8-9b41-0304ddd08ad7.html.

24. "Specifications for the United States Flag," http://www.usflag.org/flag.specs.html.

25. "State Flag—Florida Department of State," Florida Department of State, 2019, https://dos.myflorida.com/florida-facts/florida-state-symbols/state-flag/.

26. "South Carolina State House | South Carolina State Symbols," https://www.scstatehouse.gov/studentpage/coolstuff/seal.shtml.

27. FlagsImporter, "South Carolina Flag 3×5ft DuraFlag," retail website, Flags Importer.Com (blog), n.d., https://www.flagsimporter.com/catalogsearch/result/?q=south+carolina+flag+.

28. Michael Roberts, "Photos: The Eleven State Flags That Are Supposed to Be Better Than Colorado's," Westword, March 20, 2014, https://www.westword.com/news/photos-the-eleven-state-flags-that-are-supposed-to-be-better-than-colorados-5879529/2.

29. "South Carolina Flag," *Annin Flagmakers* (blog), 2020, https://www.annin.com/portfolio-archive/south-carolina-flag/.

30. "South Carolina State Flags—Nylon & Polyester—2' × 3' to 5' × 8'," www.usflagstore.com, accessed June 27, 2019, https://www.usflagstore.com/South_Carolina_State_Flag_p/south%20carolina_flag.htm.

31. Archivist ID 8, "S.C. Department of Archives and History," June 8, 2016.

32. Leslie Draffin, "S.C. Flag, Its Images Part of Fashion Statements," *Greenwood Index-Journal,* July 14, 2006, 1, 9.

33. Ibid., 9.

34. William Petroski, "South Carolina Wants to Keep 'Big Red,' Its Historic Pre-Confederate Flag an Iowa Soldier Took Home: Iowa Isn't Sure It Wants to Give It Up," *Des Moines Register,* May 9, 2018, https://www.desmoinesregister.com/story/news/politics/2018/05/09/some-black-iowan-want-voice-debate-over-south-carolina-civil-war-era-flag/551763002/.

35. Scott Malyerck, "Stop Changing SC Flag | The State," *The State* (Columbia), undated, https://www.thestate.com/opinion/letters-to-the-editor/article147462039.html.

36. "Senate Journal 3/16/2017—South Carolina Legislature Online," South Carolina General Assembly 122nd Session, 2017–2018 Journal of the Senate, March 16, 2017, https://www.scstatehouse.gov/sess122_2017-2018/sj17/20170316.htm#p6.

37. "State GOP Names Malyerck Executive Director," *Greenwood Index-Journal,* June 14, 2005.

38. Jono Miller, "How the SC State Flag Came to Be—a Bit Haphazardly," *The State* (Columbia), June 4, 2017, sec. C (Palmetto), 1C, 8C.

39. Bill 4201 to Amend Section 1–1-660 Code of the laws of South Carolina, 1976, Relating to the palmetto tree's designation as the official tree of this state, etc., Pub. L. No. 4201 (2017), https://www.scstatehouse.gov/sess122_2017-2018/pre ver/4201_20170426.htm.

40. Avery G. Wilks, "SC Has No Official State Flag Design, So Flag Makers Make It Up: That Could Change," *The State* (Columbia), January 24, 2018, https://www. thestate.com/news/politics-government/article196339369.html.

41. Cromer and Scott, Joint Resolution by Cromer, Scott, Climer, et al., Pub. L. No. S 1002 (2018).

42. Scott Malyerck, "Pickscflag (March 16 2018)," Facebook, March 16, 2018, https://www.facebook.com/Pickscflag-1394163707309170/.

43. Scott Malyerck, "Pickscflag (November 28, 2018)," Facebook, November 28, 2018, https://www.facebook.com/Pickscflag-1394163707309170/.

44. Emerson et al., *Report of the South Carolina State Flag Study Committee,* 12.

Coda

1. John Muir, *My First Summer in the Sierra* (Boston: Houghton Mifflin, 1917), http://www.gutenberg.org/files/32540/32540-h/32540-h.htm#draw7.

2. "For Want of a Nail," n.d., all nursery rhymes (website), https://allnursery rhymes.com/for-want-of-a-nail/.

3. Donald Hall and Jerry F. Butler, "Palmetto Tortoise Beetle," *Featured Creatures* (blog), August 2001, http://entnemdept.ufl.edu/creatures/orn/palms/hemi sphaerota_cyanea.htm.

Chapter 25. Lost in Space: What the Aliens Will Learn about Us

1. "NASA—Pioneer-10 and Pioneer-11," Feature Articles, Other, Mission Archives Pioneer-10 and Pioneer-11, March 2, 2012, https://www.nasa.gov/centers/ames/mis sions/archive/pioneer10-11.html.

2. Elizabeth Howell, "Pioneer 10: Greetings from Earth," Space.com, September 18, 2012, https://www.space.com/17651-pioneer-10.html.

3. "Voyager—What's on the Golden Record," https://voyager.jpl.nasa.gov/golden-record/whats-on-the-record/.

INDEX

JONO MILLER is a natural historian, environmental educator, and activist who has worked for nearly a half century to understand and protect the wild places in southwest Florida.

Jono is a both a graduate and former director of the Environmental Studies Program at New College of Florida, where he continues to volunteer. The majority of his research regarding cabbage palms was undertaken as a graduate student in the Florida Studies Program of the University of South Florida, St. Petersburg.

His writing has appeared in Susan Cerulean's *The Book of the Everglades,* the *Florida Historical Quarterly,* and the *Sarasota Herald-Tribune.*

In addition to his work on cabbage palms, Jono is known for his leadership on environmentally sensitive land protection, tree protection, barrier island issues, water management, backyard chickens, and protection of the Myakka River.

He lives in Sarasota with his wife and partner, Julie Morris.